THE
COMPASSION
MANDATE

THE
COMPASSION
MANDATE

Remaking
The European Union's
Leadership on
Farmed Animal Welfare

NEIL DULLAGHAN

BOOKS OF CHANGE ★ **SAN FRANCISCO**

TABLE OF CONTENTS

A THANK YOU

This book builds on the pioneering scholarship of researchers who painstakingly documented the early development of European farmed animal welfare policy, at a time when such work was far from a mainstream academic concern.

I am particularly indebted to Carolina Maciel and Bettina Bock, whose analysis in their book *Modern Politics in Animal Welfare?* provided crucial insights into the evolving character of European animal welfare governance. I am deeply grateful too for Michael C. Appleby's historical documentation of the EU battery cage ban and Heinzpeter Studer's similar work describing the paths taken by Switzerland and other European countries to move away from battery cage systems. These scholars took on the essential, but often undervalued, work of creating the historical record that makes this book possible. This book aims to take the baton from those scholars, carry it for a leg, and pass it on to the next generation to take it further.

I owe a profound debt of gratitude to Rethink Priorities, where I conducted the original research for the book. As a research and implementation organization dedicated to identifying and acting on opportunities to improve the lives of humans and animals, now and in the future, they provided the impact-focused and empirically rigorous foundation for this work. They have continued to provide encouragement and support for transforming that work into this present volume. And, of course, I am equally grateful to Worldshapers, who provided the crucial impetus and practical help needed to turn my research into this book.

My deep gratitude goes to my parents, Jim and Una, whose steadfast support made my academic journey from bachelor's degree to PhD possible. Their encouragement, sacrifice, and belief in the value of education provided me with the analytical tools and scholarly foundation necessary to undertake this research.

Finally, to Alex. None of this would be possible without you taking on more than your fair share while I spent evenings and weekends putting this work together, pulling me out when I was sucked in, making sure we spent time grounding ourselves by going for walks outside with our rescue dog Milo. There is no greater partner I could have, and I continue to offer my support in every way to help achieve all your goals and build a great life together.

Neil Dullaghan, July 2025

CHAPTER 1

AN INTRODUCTION

As I travelled as a citizen of the European Union (EU) across Europe for work, study, or to see friends, the agricultural sights I saw from buses, trains, planes, cars, and bikes were reassuring. Between the cities and suburbs lay neatly partitioned fields of cows and sheep grazing in the sunshine. In the terroir regions of Italy and France, the occasional chicken darted along the vines, acting as a natural insecticide. Without much pondering, one could happily conclude that the EU truly offered the standard of care for farmed animals that citizens demand and laws claim to provide.

I would ask you, though, how many times you have passed a large warehouse, just off the road, in a field maybe, and wondered: is that windowless building packed with animals confined so tightly they cannot move around or spread their limbs, sitting in their own filth, picking up diseases while their wounds fester? Maybe you have never asked yourself such a question. I know I rarely stopped to wonder. These buildings don't have signs on them saying 'Industrial Animal Farming Facility'. But that is very often exactly what they are, places where the vast majority of the EU's farmed animals spend their lives, confined not just within the limits of the building, but often inside tight stalls and crowded cages.

In 2019, a European Citizen's Initiative (ECI) brought the question hovering over such buildings to the forefront of many minds. ECIs are essentially petitions, and this one, called 'End the

Cage Age,' was signed by over 1.6 million EU citizens.[1] They were asking the European Commission to do one thing: end the extreme confinement of farmed animals. The ECI sprang from the work of animal advocacy groups determined to expose the problems with modern animal agriculture. Thanks in part to their campaigns and investigations, often undercover, it had become abundantly clear to many Europeans that the EU's agricultural system had gone wrong for animals, and that the policies aiming to protect them no longer matched social expectations. Those expectations are higher than you might think, with Members of the European Parliament (MEPs) saying they receive more letters about animal welfare than any other subject.[2]

Shockingly, even as European citizens become more aware of, and repelled by, the cruelties inflicted on animals in industrial farm systems, the number of those farms is growing. In June 2025 an investigation by the AGtivist Agency found the EU was home to at least 24,000 industrial chicken and pig farms, with some found to house more than 1.4 million chickens at any one time.[3] At a national level, the investigation showed France had the highest number of industrial chicken farms, while Spain had the highest number of industrial pig farms. Italy and Germany are in the top five EU countries for both industrial pig and chicken farms. The findings are yet another indication that the days of small family farms are long gone and that, with each passing year, a new era of super-intensive animal production beckons, with the looming spectre of twenty-six-storey buildings each packed with 1 million pigs making their way from China to a city near you.[4]

Is this consolidation of animal agriculture, into one massive cage for farmed animals, the kind of success Europeans want and hail when the EU promotes the single market? When we hear the goal is an 'ever closer union,' does this mean animals packed ever tighter? If the EU leads, is this where we want others to follow?

This book argues that there is another vision of EU leadership in agriculture, one that places compassion at the forefront. One that builds on past progress, rather than resting on the laurels of it.

THE LAWMAKING ERA

On all sides of the political spectrum and throughout its multilevel governance system, from national lawmakers to European parliamentarians, the EU pioneered a range of farmed animal protection laws in the 1980s, 1990s, and early 2000s.[5] These laws reduced extreme confinement and gave animals a little more space to behave in ways appropriate to individual species.

After these laws established the first minimum standards for farmed animal protection, there was a lull in legislative activity in the early twenty-first century as progress was delegated to a market that hides the reality of suffering behind misleading labelling and incentivizes a race to the bottom in terms of animal welfare. But citizens and animal advocates did not stop advocating for better conditions for animals. They pushed for a wave of food industry reforms that increased animal protections and went beyond existing legal standards. Then, in the late 2010s, momentum began building toward a new wave of reforms, partly in response to the 'End the Cage Age' petition.

By 2023 a major package of new and updated welfare laws was ready, intended to meet the expectations of citizens, harmonize a distorted market, and restore alignment with the latest animal welfare science.[6] The reform package would have raised welfare standards for the EU's farmed animals, including key measures to phase out extreme confinement such as cages and crates, end painful mutilations such as cutting off beaks, tails, and genitals without pain relief, set new breeding standards, improve slaughter conditions, and expand the scope of laws to cover neglected animals such as rabbits and fish. But the package was not presented as promised. Instead, a series of events – notably the war in Ukraine, rising inflation, and European elections – pushed the reforms off the agenda, leaving tens of millions of terrestrial and aquatic animals in the lurch. Suddenly, decades of progress for animal protections ran the risk not just of grinding to a halt, but of being rolled back.

Embarrassingly for the EU, this made the US look like a leader in comparison. While protections for farmed animals at the US federal level are lacking even more than in the EU, the democratic structures of the US state system responded better to their citizens' demands for farmed animal protection than the EU's. Through ballot initiatives – similar to referendums and more legally binding than the ECI – and legislative initiatives, over a dozen states, including California, were jumping ahead by banning the production (and sometimes also the sale) of meat and/or eggs from caged animals,[7] staking a claim to the moral high ground once proudly held by the EU. Adding insult to injury, while the EU has continued to intensify animal farming rather than make it more humane, it is being outpaced even in that gruesome endeavour by China. The Chinese government is investing in intensive farming of terrestrial and aquatic animals at a rapid pace, and even seeking to export its systems abroad. Watching these events unfold, it was hard not to hear echoes of Mario Draghi's warning: in yet another domain, Europe is slipping, not with a bang, but in a slow, self-inflicted agony.[8] If the EU tries to compete on cruelty it will lose, both economically and morally. In contrast, if the EU tries to compete on compassion, it stands a chance, but it needs to act fast to regain this competitive edge.

As this book goes to press, the legacies of the current 2024 to 2029 EU administration and the next (which runs from 2029 to 2034), are still unknown. Will these administrations let EU animal welfare fall further behind? Or will they get it back on track?

At its most basic, this book's central question is: will the EU seize the opportunity to show global leadership by updating farmed animal welfare standards and offering relief to tens of millions of hens, pigs, chickens, cows, fish, and other species living in outdated conditions, crammed together in environments that restrict their natural behaviours and deny them access to life beyond the bars of a cage? Or will it balk?

To answer this main question we need to ask a series of additional questions. Will the EU recognize the pace of technological progress and replace the grinding and gassing of day-old male chicks with new technology that can screen them out of the supply chain? Will hundreds of millions of fish see existing laws adapted to meet their needs during rearing, transport, and slaughter?[9] Will the EU catch up with the market and release egg-laying hens from cages? Will it protect these gains from lower-welfare imports from regions with less stringent standards? Or will it ignore the urgent needs of animals as it deals with other pressing problems?

Will commissioners, MEPs, permanent representatives, national ministers, and the teams of civil servants, economists, scientists, and legal experts that support them, recognize a new opportunity for farmed animal legislation? More importantly, certainly for the animals, this book goes beyond asking what the EU will do. It actively calls on key players, those who want the EU to once again be a global welfare leader, to step up and deliver the incentives, policies, and laws that will reduce animal suffering. Although my name is on this particular book, the call comes from a much broader collective of people: the 84 percent of European citizens who believe the welfare of farmed animals should be better protected.[10]

THE EVOLUTION OF AN INDUSTRY: HOW WE GOT HERE

For millennia, homo sapiens hunted for meat and materials, but the advent of the Neolithic Era, when crops were planted rather than gathered, led to the domestication of animals that provided food, fuel, fertilizer, clothing, and labour.[11] Killing animals for meat was often secondary, with farmers harvesting products like wool, eggs, and milk while the animals remained alive.

By the mid-twentieth century this pattern had begun to shift dramatically. Following World War II, with millions of citizens scarred by food shortages and rationing, there was a broad consensus that Europe needed to ensure such deprivations would never return. The transformation was swift and profound across the continent. The era of factory farming had begun, encouraged by government subsidies to farmers who produced more meat, dairy, and eggs using new technology.

But what started as a practical response to scarcity quickly evolved into an industrial model focused on maximizing efficiency and reducing costs. In continental Europe, the Netherlands exemplified this transformation. Between the 1950s and the 1970s Dutch livestock farming evolved rapidly from traditional to industrial systems, driven by technological developments and the desire for food security, especially after the so-called Hunger Winter of late 1944 and early 1945.[12]

By the 1950s, Sicco Mansholt, the Dutch agriculture minister, had decided the Netherlands would feed itself. Some farmers would specialize in one product and expand their business, while the government paid others to quit farming. Because space was limited in a small country, high density, high value production was prioritized, and as output successfully ramped up, agricultural exports soon became an impressive source of the country's national income.[13]

Today, the types of industrial facilities that sprouted in the Netherlands take various forms across the EU, but they share common characteristics: large numbers of animals confined in limited indoor spaces, standardized feeding and management practices, and a focus on maximizing outputs and minimizing inputs. To achieve these goals, many factory farms specialize in rearing one species, during one stage of the animals' lives – an aspect of industrialization that potentially increases the objectification of animals as units of production.

The scale of this industrialization is staggering, and it has grown exponentially over the past seventy years. Every year, the EU turns about 7 billion terrestrial vertebrate animals into meat. The vast majority of those killed are broiler chickens – 6.3 billion of them – with pigs, turkeys, ducks, sheep, cows, and goats making up the rest, in that order.[14] On top of this there are the farmed fish, of which between 670 million and 1.1 billion are killed every year,[15] plus the 390 million egg-laying hens,[16] and the almost 20 million dairy cows.[17]

In pure production terms, the results of the intensification process have been remarkable. Producers have found incredible ways to increase output while driving down costs. But the real cost of production hasn't really fallen. It has simply been passed on to the animals themselves, in ways that were kept largely hidden from consumers until non-profit organizations and animal protection charities began conducting investigations that exposed the system's inherent cruelty.

THE HIDDEN REALITY: LIFE FOR ANIMALS ON FACTORY FARMS

Most of us grew up with narratives that minimized animal suffering. We might have been told that animals must be well treated, otherwise our meat would taste bad. Or that animals have great lives in green fields and just one bad day when they go off to slaughter.

What is not mentioned in these stories are crammed, dirty warehouses, tiny cages, and bodily mutilations. Such facts would not only prompt questions about the food being served; they might raise a reasonable objection. Surely, confining so many animals, to the extent that it seriously compromises their health would be a self-limiting process. After all, sick animals are not productive. There must be a mistake.

There is no mistake. Factory farming makes animals farmed for meat and other food products very sick. For one answer as to why poor welfare and ill health are not production problems for huge farming companies, look no further than former EU agriculture commissioner, Poul Dalsager. Dalsager noted that while overcrowded cages meant that egg production per hen decreased, the maximum production per square meter of building increased, a feat achieved by packing more hens into tiny cages.[18]

Think of it like this. The Eurostar could remove all the carriage seats and pack in a lot more passengers. More people would travel, making profits rise; but the travelling experience for each passenger would be much worse. It might even make many of the travellers ill. Think of the inability to move or perform any natural train behaviours, like reading or having a coffee. If it was hot, heat stress would be a problem, to say nothing of the general discomfort and unease it would create to be so closely packed against your will. For most of the EU's farmed animals, it is much worse than this, every day of their lives.

Despite the soaring numbers, industrially farmed animals are essentially an invisible population, one that is treated more as a group of production units than sentient beings. In this invisibility, the ruthless efficiency of the production line fundamentally alters these animals in ways mostly unknown to consumers, who only see the end product sitting on clean, well-lit supermarket shelves, often adorned with misleading cartoons of bucolic scenes with smiling cows or photographs of hens roaming the fields.

This process occurs despite EU law recognizing animals as sentient beings and requiring that member states pay 'full regard' to their welfare when formulating and implementing policies.[19] The problem is that these principles come into conflict with the incentives of the market. In order to reduce production and labour costs, current EU policies allow for animals to be crowded into cages and barren pens. When it comes to slaughter, the policies are equally unhelpful. Animals are killed with so little attention to their welfare that many are boiled, butchered, and bled while still conscious.[20]

In addition to the gap between the law and the actual implementation of welfare standards, there is another issue. While genetic alterations have transformed pigs, chickens, and cows into highly efficient producers of meat, milk, and eggs that bear very little resemblance to their wild ancestors, the changes have not removed the animals' natural desires to dust, bathe, perch, wallow, root, explore, and socialize.

As I write, I am remembering a presentation I watched about breeding pigs. An image of a windswept bush appeared on the screen. Clearly, the presenter had accidentally included a shot from their personal gardening photo collection, I thought. But no, the presenter assured us that we were looking at a photo of a mother pig. Thus began a game of *Where's Wally?*, until I spotted a distinctive snout beneath the brush. Pigs naturally build nests – when not confined in coffin-sized crates, that is. My ignorance of such a basic fact about pigs humbled me, and reminded me just how far our modern system has strayed from the world these animals evolved to thrive in.

Animals that cannot perform innate natural behaviours are subject to significant stress and suffering, and each species suffers in its own way. Witnessing and reading about animal suffering is difficult: members of the European Parliament themselves refused to allow photos of standard practices on farms to be included in a photo exhibition created to show MEPs the hidden realities of everyday farmed animal suffering.[21] But it is necessary to confront and recognise animal suffering in order to make the changes required to end this suffering.

Chickens raised for meat, commonly known as 'broilers,' have been selectively bred to be incredibly hungry and grow rapidly when fed. These chicks raised for meat grow to slaughter weight by just four to five weeks of age, and do so asynchronously – growing their breast meat far faster than their hearts and lungs – so that many collapse under their own weight, suffer cardiac arrest, and die on the floors of dark barns, covered in faeces. These animals are essentially, as one NGO described it, made prisoner in their own unwieldy bodies through selective breeding.[22] Research estimates that in Europe and Canada broiler on-farm mortality rates – that is, the number of birds who die before making it to slaughter – range from 1.5% to 6.2%.[23] That may not sound high, but on farms where up to 60,000 broilers might be raised,[24] those mortality percentages suggest between 900 and 3,700 animals per flock die without ever reaching the supermarket.

To help understand the suffering better, consider the mother chickens used to produce broilers raised for meat. They too had to be selected for the trait of being incredibly hungry. But they are not fed as much as their offspring because if they were, the mothers would suffer the same health problems and die before producing more chicks. Imagine being bred to be desperately hungry, and then systematically deprived of food. That puts many broiler chickens in a lose-lose situation: either they eat as much as desired and die horrible early deaths, or they live a little longer in astonishing hunger. Even if they don't die, broilers' extreme growth rate leads to numerous health problems, including lameness.[25] Many birds spend their final days unable to walk, lying in ammonia-soaked droppings.

Chickens raised for meat are not the only ones who suffer terribly on factory farms. Egg-laying hens are often kept in cages for life, packed with five to eight other birds in spaces too small for natural movement. After about eighteen months of continuous egg production – well beyond what they'd produce in the wild – these hens are killed and discarded, far short of the ten- to fifteen-year lifespan of the junglefowl they originated from.[26]

And consider the fate of male chicks. The hens who lay eggs for supermarkets didn't sprout out of the ground or fall off a tree. They too had mothers who laid them. These mother hens, the parent stock, are forced to pump out new generations of chicks to be trapped in cages. However, only female hens can be used to lay eggs. Male chicks, useless to the egg industry and not optimized for fast growth as broilers, are typically killed a day after hatching by being ground alive or gassed.

Now let's consider pigs. These famously intelligent creatures are frequently kept in indoor facilities with no enrichment materials to distract them from the endless hours of boredom. They commonly undergo painful procedures, like having their tails cut off and their teeth clipped without anaesthesia. Pregnant pigs (sows) likely have it the worst as these mothers may be confined in stalls so narrow they cannot turn around. Partly this is to prevent them from getting into fights with other pigs, or crushing piglets in the unnatural factory environment; but it's also a great way to save resources. Pigs who don't move around much use less energy, meaning not as much feed is required.[27]

Aquatic animals in factory farming systems face their own set of severe welfare challenges. Since wild fish stocks began to collapse in the 1990s due to industrial overfishing, fish farming has become increasingly common across Europe, with producers raising various species in riverways, land-based tanks, or sea cages. If you're in the Mediterranean, these farms are mainly populated by European sea bass and Gilthead sea bream. If you're in Central or Eastern Europe, you're more likely to come across common carp. And in North and Northwestern Europe you'll find rainbow trout and Atlantic salmon, especially if you pay a visit to EU neighbours Norway and Scotland.

The welfare conditions in these operations are often abysmal. Fish are confined to swimming endlessly in circles, exposed to pathogens from being packed so tightly or suffering mass mortality events.[28] The cruelties inflicted on animals in modern fish farms are numerous and include fin clipping, overcrowding, dirty water without enough oxygen, boredom, frustration, an inability to swim freely, death by prolonged suffocation, as well as fear and stress from human handling during transport and pumping, a process that essentially involves sucking fish up a tube in order to deposit them somewhere else. Investigations have catalogued a range of fish abuses, including violent throwing and cutting fishes' gills while they are conscious.[29]

Fish slaughter practices are only vaguely regulated, and there has been almost no effort to reduce prolonged suffering. The majority of fish are killed without stunning, by slow asphyxiation in the air or in ice water, a process that can take between five and forty minutes for commonly farmed species.[30] Others die by live gutting, which can take even longer. And while some producers – usually those in wealthier European nations – have voluntarily introduced pre-slaughter stunning for fish, this remains the exception rather than the rule.

WHO CAN SOLVE THIS PROBLEM?

Faced with such terrible facts, one often looks for someone to blame. This reaction is understandable, but it would be misplaced to vilify the farmers who produce the animal products people eat and export. These human workers are just trying to make a living, responding to the incentives of the single market and government rules. Nor should we blame consumers, who are just trying to feed their families within a budget and whose choices are constrained and warped by large corporations' marketing machines. Instead of looking for someone to blame, we can look for someone to lead.

The good news is that ending this suffering is not difficult; we just need better laws, combined with proper monitoring and enforcement, and incentives for farmers. How we get those laws is a slightly lengthier story, but the following chapters aim to demonstrate that improving conditions for Europe's farmed animals is, in many cases, straightforward, and builds on a position of EU leadership taken in the 1980s and 1990s.

Nobody is in favour of animal cruelty. In fact, there has been a clear increase in public support for animal welfare legislation since industrialized farming began in the 1950s.[31] But it's easy for animal issues to become unnecessarily polarized and for people to pull on opposite ends of the rope, as if in a game of tug of war.

Instead of debating whether to have more or less legal protection, Europeans should be pulling the rope sideways,[32] discussing which combination of incentives and regulations will achieve better welfare. Instead of arguing about whether we need new laws, or whether to roll back the ones we have, EU policymakers can focus on updating existing laws to make them consistent with each other and with the latest animal welfare science.

And there is plenty to work on. Lawmakers can choose from a range of policy options already widely popular with citizens, notably the overwhelming public and market support for phasing out cages for egg-laying hens and for measures to protect EU farmers from low-welfare imports. Policymakers could also focus on introducing initiatives in a particular sequence,[33] one that would help to reduce political blowback and ensure a fairer and more just transition. For example, they should start by introducing mandatory labelling requirements and certification schemes that make lower and higher welfare production methods visible to consumers. That would allow buyers to reward the farmers shown to use more humane methods. Meanwhile, public funding can be provided to support new technologies and test husbandry methods that result in better lives for animals, earning farmers the right to use those labels.

As the market for higher welfare becomes more established and popular, there will be greater support from all stakeholders to lock in these gains, with legislation creating new minimum standards. Past successes can be repeated too. In the 2010s the EU relied on many market mechanisms and voluntary standards that led to the rise of cage-free hen farming and commitments to higher-welfare broiler chickens. Successful as these mechanisms have been, on their own they have not been enough to deliver change across the EU. Legislation is needed to underpin and support those gains and help expedite change.

WHO IS THIS BOOK FOR?

Although this book is aimed at anyone interested in animal welfare in the EU, it might be of most help to those who believe that even incremental improvements are better than none. Although it will appeal to people interested in farmed animals generally, the book will be of special interest to those concerned with the welfare of chickens and fish. These animals are the focus of this book because they make up a significant portion of the EU's farmed population and present the clearest routes for action. In choosing this focus my intention is not to detract from the suffering of other groups of animals. My reasoning is the belief that in asking the EU to prioritize welfare improvements, we should demand improvements that are both realistic and of benefit to the largest numbers of farmed animals.

The primary goal of this volume is to show how the EU became a leader in the farmed animal space, and the best ways of achieving what might still be possible during the administrations running from 2024 to 2034. Drawing on case studies of existing laws, the book lays out legal changes that could bring animals some relief in the next decade, as well as longer-term initiatives that would lay the foundations for more ambitious changes.

In the next chapter, I will look at the core species-specific directives the EU adopted between the 1980s and 2000s. These directives provide the best overview of the bloc's most and least successful attempts to protect as many of the numerous animals in its care as possible. Chapter 3 offers a brief history of the EU's specific laws; Chapter 4 describes the various challenges encountered in the 2010s, as the EU tried to bring the words of those laws to life, and showcases the best practice (the 'dos and don'ts') for any future legislation.

Chapter 5 and Chapter 6 look at progress toward getting egg-laying hens out of cages and why an EU cage phase-out is the next obvious policy choice, along with ensuring the domestic market is protected from imports produced with lower welfare standards. Chapter 7 suggests a possible path toward protecting aquatic animals for the 2029 to 2034 administration. Finally, Chapter 8 looks at what the benefits might be, for animals and the EU itself, if we implemented all these proposed measures.

This book is therefore both a guide to stoking policy changes and a pragmatic appeal to those in positions of power to continue making progress for animals. It argues, essentially, that for the EU to be a leader on the global stage, it must not compete on purely narrow economic grounds. It must be able to outcompete others on moral grounds and show that the single market can deliver real prosperity and flourishing for both humans and the non-human animals we depend on.

CHAPTER 2

THE BIRTH OF FARMED ANIMAL PROTECTION IN THE EU

The 1964 release of a factory farming exposé called *Animal Machines* is often seen as initiating European governments' active engagement in creating laws to protect farmed animals.[34] The book is credited with revealing to the British public the realities of the newly emerging system of industrial animal farming.

The author, Ruth Harrison, wrote the book after a campaign group called Crusade Against All Cruelty to Animals pushed a leaflet through her door sometime around 1960.[35] In doing so, 'they spurred to action a politically compatible and well-connected Quaker, who knew the power of civic activism,' writes Claas Kirchhelle, who has documented Harrison's work.[36] *Animal Machines*, says Kirchhelle, contained 'a series of disturbing images and descriptions of animals' living conditions under intensive housing and slaughtering practices in abattoirs.' The book's foreword was written by the famous author of the groundbreaking American environmentalist volume *Silent Spring*, Rachel Carson[37] – whose name appeared more prominently on the cover of *Animal Machines* than Harrison's. Carson's endorsement helped to turn Harrison's book into an instant bestseller.

The impact of *Animal Machines* reverberated far beyond individual readers. Six weeks after it hit the shelves, the British government appointed a committee, chaired by Roger Brambell, tasked with examining 'the conditions in which livestock are kept under intensive husbandry and to advise whether standards ought to be set in the interest of their welfare, and if so what they should be.'[38] A year and a half later, the Committee's report laid the foundations of an ethical and scientific framework for farmed animal welfare and introduced the idea of five animal freedoms: 1) freedom from hunger and thirst; 2) freedom from discomfort; 3) freedom from pain, injury, or disease; 4) freedom from fear and distress; and 5) freedom to express normal behaviours.[39]

Next came Sweden. In 1968, a series of critical articles by veterinary surgeon Kristina Forslund and the much beloved author Astrid Lindgren,[40] famous for her Pippi Longstocking books, appeared in the Swedish national newspapers, *Expressen* and *Dagens Nyheter.*[41]

The articles chided the Swedish agriculture minister for the ways animals were being raised on industrialized farms. Radio programmes and debates in the Swedish parliament followed, and the Swedish Church issued a study on animal ethics.[42] At the same time, scientific research was finding that new labour- and space-saving technologies for animal production were detrimental to animal wellbeing.[43] This kicked off decades of pressure for reform that culminated in 1988, when the Swedish agriculture minister presented an animal welfare law to Lindgren on her birthday as a kind of honorary tribute.[44] [45]

The trend had been set. During the final decades of the 20th century, several European countries created governmental bodies to work on farmed animal protection policies. The Farm Animal Welfare Council was established in the UK in 1979,[46] the Tierschutzkommission in Germany in 1987,[47] and in 1993 the Dutch Raad voor Dierenaangelegenheden.[48] As we will see throughout this book, in the first era of farmed animal protection legislation, a handful of countries in Northwestern Europe were the first to take the lead due to a combination of social, economic, and food safety forces.

Continent-wide initiatives were emerging too, with the Council of Europe adopting the Convention on Animals in Transport in 1969, and then the European Convention for the Protection of Animals Kept for Farming Purposes in 1976. Countries that were members of both the Council of Europe and the European Economic Community went on to ratify the Convention as legally binding,[49] although countries with a higher share of their national workforce employed in agriculture (like Greece and Spain) were slower to make that move than countries that had more urbanized workforces (like Belgium and the Netherlands).[50]

Lobby groups were being formed too. In 1980, Britain's Royal Society for the Prevention of Cruelty to Animals (RSPCA) founded an EU lobby called the Eurogroup for Animal Welfare (now Eurogroup for Animals), along with five other EU member organizations: Dierenbescherming in the Netherlands, Deutscher Tierschutzbund in Germany, Dyrenes Beskyttelse in Denmark, la Société Protectrice des Animaux in France, and Lëtzebuerger Déiereschutzliga in Luxembourg.[51] The Eurogroup for Animals lobbies at the EU level on behalf of its members and acts as the secretariat for the European Parliament's Intergroup on the Welfare and Conservation of Animals, formed in 1983.[52]

A CRISIS POINT IS REACHED

The outbreak of several farmed animal epidemics in the 1990s generated widespread public concern about the negative effects of intensive farming, and undermined consumer trust in both food safety and the government and public systems charged with controlling that safety.[53] The worst of these epidemics was the UK's 'mad cow' disease, an illness in cows more correctly known as bovine spongiform encephalopathy or BSE. The outbreak has been attributed to the practice of feeding dead cows to other cows, usually in the form of meat and bone meal (MBM).[54] After human deaths caused by variant Creutzfeldt-Jacob disease were linked to eating BSE-infected meat, a public panic ensued.[55] Over 4 million cows were killed in an effort to constrain the spread of the disease. The crisis also led to the UK instituting a ban on the use of MBM in 1988,[56] followed by the EU doing the same by 1994.[57] Other similar crises around salmonella in eggs, antibiotic resistance, swine fever, and bird flu led to a number of reforms of animal farming practices and a growing acceptance of animal welfare concerns as an area government should act on, though often framed in terms of human food safety.

Together, the BSE crisis and strong northern European public and political interest in animal welfare presented an opportunity for the European Commission, the executive body of the emerging European Union, to resolve legal distortions in the common market by proposing EU-wide animal welfare laws. Throughout the 1980s and 1990s, the Commission proposed a number of species-specific policies, to protect calves, pigs, and egg-laying hens. These policies were eventually passed into law, albeit after negotiations and amendments by EU countries to grant lengthy transition periods for producers.

In 1992, a formal declaration on animal protection was annexed to the EU's core governing document at the time, the Maastricht Treaty, which called on EU institutions and countries to 'pay full regard' to their welfare requirements. The annex itself was the result of years of campaigning by the animal welfare charity Compassion in World Farming (CIWF), founded in Britain in 1967 to promote the welfare of farmed animals.[58] This was followed in 1997 by a protocol that was annexed to the subsequent Amsterdam Treaty, recognizing animals as sentient beings and stating that, 'in formulating and implementing the Community's agriculture, transport, internal market and research policies, the … [EU] shall pay full regard to the welfare requirements of animals.'[59] Animal sentience was initially recognized as a non-binding Treaty Declaration before being given legal status in the subsequent Treaty of Amsterdam. Finally, in 2007, animal sentience was recognized as an article in the Lisbon Treaty, giving it additional weight within the EU.[60]

THE PACE SLOWS

Since this last legal landmark in 2007, the pace has slowed. As of writing this book in 2025, the most recent farmed animal welfare law passed was in 2009, covering the protection of animals at the time of killing.[61] This was technically a new law, but in reality it was a revision of an older law from 1993.[62] While the Animal Health Law and the Official Controls Regulation were adopted in 2016 and 2017 respectively,[63] these were more focused on disease risks and food safety than animal welfare. The last species-specific EU legislation was the broiler directive, passed in 2007.[64] For the most part, over the past two decades, instead of progress there has been controversy, with major problems emerging in the 2010s around non-compliance with laying hen and sow laws, despite producers having had more than ten years to prepare – an issue we will return to in Chapter 4.

As the topic of animal welfare crept down the legislative agenda, mostly relegated to platforms, strategies, and audits, there was one attempt to improve matters. As part of the 2012 to 2015 EU Strategy for the Protection and Welfare of Animals,[65] the EU assessed the feasibility of introducing a simplified legislative framework aimed at improving the application and enforcement of welfare laws. But the initiative was not completed as intended. Instead, the EU said the evaluation's findings would feed into the farmed animal welfare law fitness check,[66] which began in May 2020,[67] and would be considered for future actions in the animal welfare area.

REASONS FOR THE PAUSE

The pause on any major farmed animal welfare legislation – which effectively ran from 2007 up to about 2020, when animal welfare reform discussions began again as part of the European Green Deal and Farm to Fork Strategy – can be attributed to a range of factors.

First, this was a period of relatively rapid expansion of the EU. Between 1995 and 2007 the Union grew from fifteen to twenty-seven countries. The new member states were mainly Central and Eastern European countries whose governments, broadly speaking, did not see animal welfare as a priority compared to economic and infrastructure development. These were states that had transitioned from socialist regimes that lacked significant animal welfare laws. These regimes had consolidated animal agriculture into state or collective indoor farms and banned animal protection societies, while students in veterinary and agricultural institutes were discouraged from discussing welfare and the state-run media rarely covered the issue.[68] The accession of a number of post-socialist states in the late twentieth and early twenty-first centuries fundamentally changed the prospects for legislative progress on animal welfare in the EU, raising the bar needed for a majority Council vote and shifting the balance of power toward those that favoured the status quo.

Second, there was growing competition from animal producers outside the EU who were offering cheaper, lower-welfare animal protein imports. This could have provided an incentive for EU leaders to continue advancing standards for farmed animals, and to apply those standards to imports in order to protect domestic producers and reinforce the understanding that EU products were held to higher welfare standards. Instead, EU decision-makers at the time missed this opportunity to show leadership, and let the smaller and higher-welfare farmers lose ground to corporations and megafarms trying to compete with producers abroad.

Third, during this same period, governmental policies were shifting from regulatory to voluntary approaches in cooperation with the private sector.[69] The European Commission hoped that better consumer welfare information would enable farmers to ensure higher standards of animal welfare without risking income loss,[70] and without the need to pass new laws. To that end, private and national food labels were developed which reflected the different levels of animal welfare which had been observed in the production of individual products.

The problem was that many industry-led guidelines were simply used as cover to continue business as usual. The concept of 'greenwashing' was extended to animal agriculture. The phrase 'humane washing' was used by many to describe a process that saw animal-friendly terms slapped onto labels without meaningful change on the ground. Thus it fell to civil society to put pressure on corporate producers and retailers to actually improve their animal welfare standards,[71] especially for egg-laying hens and broiler chickens.[72] For example, in 2007, the Dutch animal welfare NGO Dierenbescherming introduced a scheme called Beter-Leven or Better Life,[73] which awarded one, two, or three stars to animal products. One star indicated that basic improvements had been made to animal welfare. Three stars meant the meat, eggs, and dairy met even higher standards. Another Dutch animal welfare NGO, Wakker Dier, ran a campaign raising awareness about the plight of broiler chickens bred to grow fatally fast, calling them *plofkip* or 'exploding chickens.' The growing salience of the issue and the disconnect between the reality on the ground and the welfare scheme created by the retailers themselves led to a sea change.[74] By 2021, all Dutch supermarkets, led by Albert Heijn, promised to ensure all their chickens carried at least one star from 2023 onward.[75] But as positive as these examples were, they were more the exception than the rule. As of 2025, the vast majority of farmed animals in the EU are still held in extreme confinement.

In addition to the factors cited above, from 2007 to 2025 the EU has had to deal with a series of wider pressures: the effects of the global financial crisis, its own Eurozone debt crisis, the immigration crisis, the fallout from Brexit, and rising populism. All of these have contributed to an environment in which animals needed to fight even harder to get attention. However, before that pace slowed, the EU did take bold steps to pass the first continent-wide species-specific protections for farmed animals and create its reputation as a leader in animal welfare.

CHAPTER 3

WHEN THE EU LED THE WORLD: A BRIEF HISTORY OF SPECIES- SPECIFIC LAWS

Between the 1980s and 2000s, the European Union achieved something unprecedented in global governance: the world's first comprehensive international regulations on industrial farming practices. These landmark directives – banning battery cages for hens, phasing out sow stalls for pregnant pigs, restricting veal crates, and setting minimum stocking-density standards for broiler chickens – emerged from a complex interplay of national fragmentation, consumer pressure, scientific evidence, and strategic leadership.

While each case followed a similar pattern – early-adopting countries creating competitive disadvantages that demanded EU-wide harmonization – the political dynamics and ultimate outcomes varied significantly. The following case studies, which summarize more detailed research,[76] reveal both the possibilities and constraints of using European integration to advance animal welfare, demonstrating how the EU's unique institutional structure enabled ethical progress that individual member states could never have achieved alone, while also showing how industry resistance and political compromise shaped the boundaries of what was possible.

MORE SPACE FOR EGG-LAYING HENS

The earliest stirrings of legislative action against battery cages for egg-laying hens in Europe began not at the EU level, but within individual European nations, most notably Denmark, Sweden, Switzerland, and the United Kingdom.

Denmark's 1950 Protection of Animals Act was initially interpreted as banning battery cages, which often gave hens as little as 450 cm² space – less than the size of an A4 sheet of paper or an iPad.[77] The Act was driven by both animal welfare concerns and fears that industrial production would eliminate small farms.[78] It created immediate market distortions. As a result of the ban, Danish companies simply moved production across the border to Germany, while cheaper imports flooded the domestic market. By the late 1970s, egg production from hens in cages (which was illegal) may have comprised up to half of Denmark's total egg output, forcing the government to abandon the ban in 1979 in favour of regulations requiring a 600 cm² minimum cage space.[79] By comparison, in 2025, current EU law says that in enriched cages each hen must have at least 750 cm² of cage area – still only slightly bigger than an A4 sheet of paper to spend their entire lives on.[80]

In 1965, a UK government committee on the welfare of farmed animals presented its findings in the Report of the Technical Committee to Inquire into the Welfare of Animals Kept under Intensive Livestock Husbandry Systems – widely known as the Brambell Report.[81] Although the Committee did not push for a return to traditional free-range hen farming, it did recommend at least 730 cm² floor space for hens.[82] These recommendations were not put into place in the UK at that time. However, the recommendation was a signal of what was to come.

In 1976, the Council of Europe produced the relatively vague Convention on the Protection of Animals Kept for Farming Purposes,[83] and in 1979 the EU financed background scientific work on poultry welfare, declaring that it intended to end authorization for the keeping of laying hens in cages. But in 1980 the European

Commission instead decided to pass a directive for the protection of laying hens raised in battery cages, meaning that they had abandoned the idea of a total ban.[84] With the Commission failing to act, in 1981, the UK considered calling for an end to EU battery cages by 1986, with a five-year transition period and minimum space requirements of 550 cm² to 750 cm² in the meantime. At this point, about 80% of Europe's flock, based on 1979 data, were kept in cages,[85] and about half had less than 450 cm² to live in.[86]

The Commission's 1981 proposal –requiring 500 cm² minimum space with a 1995 phase-out deadline for smaller cages[87] – revealed the tension between welfare and economics. In fact, the Economic and Social Committee of the Council adopted an opinion placing the protection of animals below that of humans, stating the belief that 'the protection of animals is not one of the fundamental objectives of the Treaty of Rome.'[88]

Despite the European Parliament calling for a nearer deadline for phasing out cages by 1990, political momentum stalled through the early 1980s. Southern European countries, where most hens had less than 450 cm² space, resisted restrictions. The final directive was adopted in 1986, the year by which the UK was originally hoping to phase out cages, and it represented a compromise: 450 cm² minimum space with a 1995 implementationdeadline.[89]

It later emerged that many countries failed to implement the changes proposed in the directive.[90] Despite the problems, however, the directive marked the EU's first concrete step toward transforming industrial egg production for the benefit of hens, establishing the principle that animal welfare could justify market harmonization, even when implementation remained uneven across member states. Market and technological forces soon left the directive outdated, however.

NATIONS MOVE AHEAD

Without having to find a compromise with neighbours who had lower standards and less ambition, in non-EU Switzerland the transition away from battery cages was achieved more quickly, by both laws and private retailer initiatives.

Switzerland pursued a different path from the EU, with animal advocates securing an 85% popular vote for cage-free hen housing in Zurich in 1969, after twice failing to pass a national animal welfare act. Despite campaigners collecting 400,000 signatures on a petition for a national cage ban, backed up by scientific recommendations for prohibition, the federal government yielded to industry pressure and resisted the ban. In 1978 the Swiss Animal Welfare Act was passed – without a cage ban. A few years later, after a draft of regulations that would put the Welfare Act into practice leaked to the press, more petitions demanding a cage ban were signed and the law was eventually amended to require 800 cm^2 for each bird by 1991, plus nests and perches – standards that would eventually make battery cage authorization effectively impossible.[91]

When the Swiss Federal Veterinary Office withdrew authorization for all new cage systems in 1986, the country's two largest supermarket chains, Migros and Coop– which controlled over half of domestic egg sales – announced they would stop purchasing battery eggs from 1989 onward. This retailer-led boycott drove Swiss farmers to abandon cages entirely, with over 80% of battery farms placing orders for alternative systems by 1990.[92] This radical transformation was an early sign that battery cages had an expiry date, although the EU would be slow to follow Switzerland's lead.

Even before it joined the EU, Sweden had already been moving things forward. Around 1985 the country pioneered the 'enriched cage' concept. These novel cages reduced hen injuries while providing more space and opportunities for natural behaviours.[93] They were implemented during a period of research projects on alternative systems for laying hens that had been initiated by grants from animal welfare associations and funds from the Swedish Board of Agriculture and the Swedish University of Agricultural Sciences. This was also a period of major activity by the country's animal protection organizations, many of which switched their focus away from animals used in experiments to those on farms,[94] revealing to the public the harsh realities of caged systems. Adding to the momentum were the humorous Astrid Lindgren articles criticizing the agriculture minister about farmed animal conditions. The famous author's writing on animal suffering generated huge public support as well as legislative change.

Sweden's 1988 Farm Animal Protection Act promised a complete cage phase-out by 1999.[95] However, implementation proved difficult because early cage-free systems showed high mortality rates. The egg industry successfully lobbied for delays,[96] and the word 'cage' disappeared from the Act's paragraph text, which meant enriched cages became a permitted alternative. However, enriched cages had to undergo new technology testing before they were allowed to be used more generally.[97] In 1999, instead of phasing out cages altogether, Sweden's Ministry of Agriculture allocated SEK2.5 million to informing consumers about egg production and various hen-keeping systems. Among other things, the move aimed to motivate consumers to make choices that promoted good animal care.[98] This is an example of leaders lowering ambitions for animals, but mediating the loss by offering market support for higher welfare options.

The 1990s saw explosive growth in cage-free egg markets across wealthy European regions. By 1998, 40% of Dutch retail eggs were cage free, while leading UK supermarkets reported 25% to 52% cage-free sales, with major British retailer Marks & Spencer achieving 100% free-range eggs.[99] However, it's important to note that the UK's success was not linear. In 1951, free-range eggs accounted for 80% of all eggs produced. That had fallen to 1% by 1980.[100] In that same year, 85% of UK hens were in cages despite the fact that 89% of Britons considered them to be cruel.[101] From this 1981 nadir, free-range sales rose to 60% of the 11.3 billion eggs produced in 2022 and, by 2024, free-range eggs accounted for 72% of sales.[102]

In Germany, demand for alternative eggs was so strong that by 1999, 60% had to be imported due to the lack of domestic cage-free supply.[103] Even partial advances created momentum. For example, in 1997 Swedish retailer Hemköp stopped selling cage eggs. Although other supermarkets didn't immediately follow suit, battery-cage eggs were facing waning customer interest, making them harder for stores to move, even if most chains hadn't yet formally banned them.[104]

Similar patterns were found in other European countries. In Austria, federal states began banning cages from the mid-1990s, with Vienna prohibiting them from 1994,[105] and others following through the 2000s.[106] Finland's 1996 animal welfare legislation initially banned cages entirely,[107] although this was later softened. The situation was not helped by Swedish farmers selling tens of thousands of old cages, enough to hold 300,000 hens,[108] to their Finnish counterparts after Sweden introduced its cage restrictions. In the words of one Finnish report: 'Since Sweden banned the use of traditional cages, the import of used cages from Sweden to Finland has been quite lively.'[109] This is an example of the sometimes unpredictable consequences of individual nations implementing animal welfare legislation without wider EU guidelines. The pressure for greater uniformity was gaining momentum, however. By the late 1990s, Sweden, Germany, and Denmark already required minimum cage sizes that exceeded EU requirements under some conditions,[110] and national governments in the UK and the Netherlands were pushing for changes to the EU standards.[111]

THE EU CATCHES UP

In October 1996, the EU's Scientific Veterinary Committee adopted an opinion listing the welfare benefits and deficiencies of cages and non-cage systems. The opinion drew on the consensus of scientific studies to suggest that 'hens are prepared to work to increase their space up to at least 775 cm^2 per bird,' but that as much as 1000 cm² surface area per animal 'allows the bird to ex-press a large variety of behaviours.'[112]

Faced with new scientific evidence, growing consumer pres-sure, and an increasing patchwork of national restrictions, the Eu-ropean Commission proposed new legislation in March 1998 re-quiring 800 cm² minimum cage space, nearly double the existing standard. The proposal set ambitious deadlines: new installations had to comply by 1999, while existing systems faced phase-out by January 2009. An interim increase to 550 cm² would take effect in 2004. At the same time, the Commission explicitly acknowledged that 'alternative housing systems also still have some disadvantag-es which have not been solved yet entirely, and therefore it is too early to ban battery cages.'[113] This pragmatic position reflected sci-entific uncertainty about cage-free systems' welfare benefits, even as political pressure mounted for more radical change.

How did member states feel about this proposition? Luckily for us, there is a dataset on decision-making in the European Union on 331 controversial issues raised by 125 legislative proposals in-troduced between 1996 and 2008.[114] Using semi-structured inter-views, the study identifies the policy alternatives most favoured by each of the main political actors. It shows that prohibiting certain cage types was strongly supported by the European Parliament, Austria, Denmark, Finland, and the UK, while the Commission, the Netherlands, Sweden, and Germany partially supported the move; most of these political actors favoured at least 800 cm² of cage area. Austria, Finland, and Germany preferred a 2015 dead-line and the others a 2012 deadline. Belgium, France, Greece, Ire-land, Italy, Portugal, and Spain all favoured the status quo.

Germany's position as EU Council president in the first half of 1999 facilitated movement. Aided by having an agriculture minister from the Green Party, the country pushed for detailed enriched cage specifications, rather than an immediate prohibi-tion of all cages.[115] This technical approach helped resolve what had become a deadlock between the extremes of those who wanted a total ban and those who opposed it.

A decisive moment came in May 1999 when Italian animal advocate Adolfo Sansolini began a hunger strike,[116] forcing the Italian government to reverse its opposition to reform within 40 hours. Adding to the change in momentum, in early July the German government lost a legal attempt, against its own laws, to maintain the use of battery cages,[117] and just days later the EU adopted its own partial cage ban.[118] Although this brought those favouring more space for hens closer to a majority, France, Portugal, and Greece still resisted until last-minute concessions secured their support.[119]

The final directive reflected extensive compromise. The implementation deadline was pushed back to January 2012 for existing systems, and 2003 for new installations. Space requirements were reduced from 800 cm^2 to 750 cm^2, the same standard the UK had originally sought in 1981. Most significantly, enriched cages were explicitly permitted as an alternative to cage-free systems. This outcome demonstrated both the EU's capacity for incremental welfare progress and the constraints imposed by economic competition concerns.

MORE SPACE FOR PREGNANT PIGS

Sweden again pioneered European farmed animal welfare reform, this time for sows. Swedish producers had begun experimenting with group housing systems after a 1971 ban on tethers – devices used to restrict individual sows, typically by tying them in place by the neck or body. Scientists were looking to develop alternative systems based on ethological research into sows' maternal behaviours. This work was further developed following research from 1973 that documented the adverse consequences of intensive confinement systems for pigs in early pregnancy, known as sow stalls or gestation crates.[120] (These crates are different from the farrowing crates pigs are also confined in, from just before they give birth to many weeks after.)

With Astrid Lindgren's celebrity advocacy generating public pressure, Sweden's 1988 Farm Animal Protection Act mandated a complete phase-out of sow stalls by 1994, a year before the country would join the nascent EU.[121] However, Sweden's unilateral actions contributed to immediate competitive disadvantages after its EU accession in 1995. Swedish pig self-sufficiency dropped from nearly 100% to 68% as cheaper imports flooded in from Germany and Denmark, where production costs remained lower under intensive systems.[122] This pattern would repeat itself in other early-adopting countries, demonstrating the economic pressures that drove calls for EU-wide harmonization.

The UK became the first major EU country to legislate against sow stalls, pushed by mounting scientific evidence and a remarkable grassroots campaign. In 1988, the UK Farmed Animal Welfare Council published an assessment that found it was possible to ban new installations of stalls and tethers, but passed onto the government responsibility for deciding when the alternatives would be viable and sufficiently free of welfare problems.

The farmed animal advocacy organization CIWF convinced Conservative MP Sir Richard Body to propose legislation banning narrow stalls and chains for pregnant sows. The campaign, enhanced by celebrity support from actor Joanna Lumley, as well as concerns about food safety in the wake of the BSE crisis, generated massive public response, with one MP reportedly receiving 'more letters on pigs than on the Gulf War,' (in which the UK participated from 1990 to 1991).[123] Despite cross-party parliamentary support, the Conservative government delayed Sir Body's private member's bill by introducing its own regulations in January 1991.[124] These regulations banned any new sow stall installations immediately and required the phase-out of existing systems by December 1998. As a result, British pig production fell 40% between 1997 and 2007, while imports from Denmark and the Netherlands surged.[125] The UK's National Pig Association, initially opposed to the ban, soon joined the National Farmers' Union in lobbying for EU action to prevent unfair competition from cheaper imports.[126]

The European Parliament had called for ending sow stalls as early as 1987, but the Commission's 1989 proposal for the welfare of pigs avoided recommending general bans.[127] The Economic and Social Committee dismissed welfare concerns as exaggerated, arguing that group housing research wasn't sufficiently advanced for definitive recommendations.[128] Despite continued attempts in early 1990 by Members of the European Parliament to restrict the use of sow stalls, the final law in 1991 only agreed to phase out pig tethers.[129] This partial measure created an unintended consequence that would drive the need for further reform. Rather than converting to more humane group housing or free-range systems, many producers replaced tethers with sow stalls.[130]

But the directive also required a scientific review by 1997 to assess welfare implications and propose further reforms. When it arrived, the 1997 EU Scientific Committee report provided compelling evidence for group housing, noting that sows 'should preferably be kept in groups,' citing multiple studies showing the health benefits of loose housing systems.[131] By this time, several countries had moved beyond the EU's limited requirements. Sweden and the UK had implemented substantial group housing,[132] while the Netherlands mandated group housing beyond the first week of pregnancy for new systems from 1998,[133] following a devastating swine fever outbreak that heightened public awareness of intensive farming. Similarly, political pressure and consumer demands led to Denmark requiring group housing of sows after four weeks –from 1999 for new installations with full compliance by 2012.[134] All these national measures reflected both scientific consensus and growing consumer pressure.

In 2000, NGOs CIWF and the European Coalition for Farm Animals called on the European Commission and EU agriculture ministers to phase out sow stalls altogether. The organizations collected 660,000 signatures in a petition and produced a report titled *The Welfare of Europe's Sows in Close Confinement Stalls*.[135] The report included a survey of 11 European pig welfare experts who ranked stall and tether systems at the bottom for sow welfare.[136]

The following year, under Sweden's rotating EU presidency in January 2001, the Commission proposed phasing out sow stalls after four weeks by 2012. Five member states had already adopted restrictions beyond EU requirements, creating market pressure for harmonization. That pressure was highlighted by the Commission, which noted that while '[m]ost consumers are unlikely to be willing to pay more for pigmeat that is produced under conditions which they perceive to be better with respect to animal welfare,'[137] the economic threat to European producers from regulatory fragmentation justified action.

The final reforms adopted extended the phase-out deadline to January 2013.[138] This represented a significant advance over the status quo but also had some adverse consequences, such as the Dutch government shifting their own deadline to phase out existing sow stalls from 2008 to 2013 to match the EU standard, meaning many more pregnant pigs could be confined for years longer. For better or worse, this is another example of the EU's non-linear progress on protecting farmed animals. Instead of a straight line, we zig and zag, but always, hopefully, toward a better future.

MORE SPACE FOR CALVES

In the 1970s and 1980s research results provided evidence of serious welfare problems resulting from the extreme confinement of calves raised for veal. Progress to remedy that situation began with a cascade of unilateral national actions that created growing market distortions.

In 1989, Sweden initiated a ten-year phase-out of narrow confinement for calves raised for white veal production, as part of wider reforms to farmed animal husbandry.[139] Neighbouring Finland would later ban crates entirely in 1996, following EU acces-sion.[140] Meanwhile, in the UK, CIWF founder Peter Roberts had launched one of the country's largest consumer boycotts, target-ing white veal production in narrow crates. His campaign proved so effective that by 1987 only eight veal crate farms remained in the UK. The boycott led to a 1990 government ban on crates for calves beyond eight weeks of age.[141] The European Parliament is-sued a resolution in 1987 arguing that 'calves should not be de-prived of social contact with other calves after six weeks of age.'[142] UK MPs began helping the Commission with the preparatory work and suggested support for an EU community-wide measure as soon as it was tabled.[143]

In 1989, the Commission initially proposed a ten-year transi-tion after which calves 'shall not be confined in individual boxes or by tethering in stalls after the age of 8 weeks.'[144] Industry resis-tance proved formidable, however, with the Economic and Social Committee warning against 'systematic and hasty implementa-tion' of welfare improvements.[145] By November 1991, the directive had been severely weakened, merely to require perforated walls and minimum crate widths, rather than limiting calves' time in individual confinement.[146]

Despite the weakened requirements, group housing was eco-nomically attractive and market dynamics were changing across Northern Europe. In the Netherlands, the EU's second-largest veal producer, Dutch supermarket chain Albert Heijn refused to sell crated veal, sparking a supply-chain revolution. Group hous-ing increased dramatically from 4.7% in 1985 to 22.9% by 1995, eventually reaching 60% by 2001, as Dutch farmers were 'fran-tically changing to group housing' because welfare standards proved economically viable.[147]

This market shift carried enormous implications since the Netherlands exported 95% of its veal products, supplying 45.6% of German, 32.6% of Italian, and 16.6% of French consumption re-spectively. Consumer surveys revealed that 64% of French and 45%

of Italian consumers were willing to pay premiums for higher welfare veal.[148] But domestic producers failed to respond to this market signal. French veal consumption fell by 25% between 1988 and 1994, while production declined in both France and Italy, where group housing remained limited. With consumer concerns driving rapid change in welfare standards, countries maintaining crate systems found themselves increasingly isolated from export markets and traditional producers struggled with declining demand.[149]

The UK's ban on veal crates immediately generated perverse outcomes. 40% of calves[150] born in the UK were exported live to continental member states to be raised in crates, slaughtered, and reimported as veal meat. Consequently, domestic farmers were being threatened by low-welfare imports, and the export-reimport cycle revealed the absurdity of maintaining different national rules within a supposedly integrated market.

The situation became untenable by 1995, when mass protests erupted throughout the UK against live calf exports.[151] The RSPCA and CIWF had presented a legal challenge to the government's refusal to ban exports,[152] highlighting a fundamental contradiction: British taxpayers supported higher welfare standards, while British calves were shipped abroad for practices banned domestically. A cross-party House of Commons select committee strongly supported a stand made by William Waldegrave, state secretary for agriculture, in seeking a Europe-wide ban on veal crates in response to protests over the calf export trade,[153] and a junior agriculture minister toured European capitals to press the UK's view.[154]

In the face of these challenges, the Commission faced a stark choice between allowing continued market fragmentation or developing EU-wide standards that could resolve both welfare and competitive concerns. Several factors made comprehensive action necessary as well as politically feasible. The UK had assembled support from eight other countries,[155] creating a potentially winning coalition. More importantly, perhaps, the Netherlands' market success demonstrated that group housing was commercially viable for major exporters.

The Commission accelerated its scientific review,[156] likely recognizing that credible evidence would be essential for overcoming remaining industry resistance. When the November 1995 EU Scientific Committee report confirmed that group housing significantly reduced welfare problems, compared to individual pens, it provided the technical justification needed for comprehensive reform.[157] The report was particularly important because it addressed industry claims about disease and mortality risks in group systems. University of Cambridge Professor Donald Broom, who chaired the expert working group, later emphasized that 'scientific evidence led to the Directive banning veal crates,' noting that politicians 'could not hold out against the comprehensive scientific evidence of poor welfare in crates, the availability of commercially viable alternatives and the support of European consumers for a ban.'[158]

The Commission's January 1996 proposal represented strategic leadership in resolving a crisis of regulatory fragmentation. By requiring group housing after eight weeks for new holdings to be implemented from 1998, and phasing out existing holdings by 2008, the proposal addressed both immediate competitive concerns and long-term welfare objectives.[159] The timing proved crucial. Launched during the Italian Council Presidency, the proposal demonstrated that even traditional veal-producing countries recognized the need for change. Despite initial French resistance, the Commission's approach of providing adequate transition periods while maintaining clear deadlines built the necessary support. The final text, which passed into law in January 1997, even set an earlier deadline of 2006 to phase out existing systems, bringing the benefit to even more calves.[160]

MORE SPACE FOR BROILER CHICKENS

Chickens raised for meat are not typically kept in metal cages, but rather caged in their own bodies which have been selected to grow incredibly quickly.[161] Given the fixed size of the buildings in which broiler chickens are typically raised, the faster the birds grow, the less space each individual has amongst their ballooning peers. The space allowed for each bird, expressed in kilograms per square metre, is described in regulations as a 'stocking density'. The higher the weight allowance, the less space for each growing bird.

By the early 2000s, a complex patchwork of national broiler welfare standards had emerged across the EU, creating the familiar pattern of market distortions and competitive disadvantages. For example, Austria's federal states had established maximum stocking density rates of 30 kg/m²,[162] while Denmark implemented legislation reducing density from 44 kg/m² to 40 kg/m² by 2006.[163] Germany set maximum limits of between 30 kg/m² and 39 kg/m² following university research concluding that broilers prefer stocking densities lower than 42 kg/m².[164] Sweden set an upper limit of 36 kg/m²,[165] while the UK's Farm Animal Welfare Council recommended 34 kg/m² maximum in 1992.[166] Voluntary retailer initiatives were emerging too, such as France's Label Rouge scheme and the UK's Freedom Food (now RSPCA Assured), which set 25 kg/m² and 30 kg/m² limits respectively,[167] though uptake remained limited.

The Commission's response to this patchwork of regulations began with gathering scientific evidence. In March 2000, the EU Scientific Committee produced a report called The Welfare of Chickens Kept for Meat Production. It noted welfare problems when densities exceeded 30 kg/m². Crucially, the report stressed that 'an increasing number of consumers are expressing concern about the welfare of chickens since this is being reflected in the welfare standards during production by some food retailers,' a reference to a small group of voluntary schemes with lower stocking density limits.[168]

In the UK, the RSPCA produced a report in 2001 called *Behind Closed Doors*,[169] while CIWF launched legal action against the UK government for inadequate regulations and published its own report, *The Welfare of Broiler Chickens in the European Union*, in 2005.[170] The European Commission would later cite these two British reports as evidence of the need for EU-wide action on chicken welfare. By July 2003 even the British Poultry Council welcomed EU-level work, arguing that in 'a single market single market rules should apply.'[171] This suggested that the industry finally recognized that harmonized standards were preferable to continued fragmentation.

From January 2002 to September 2003, the Commission hosted seven working groups with national farm inspectors from Sweden, Denmark, Netherlands, France, Spain, Germany, Austria, and the UK. This technical dialogue built expertise across member states and demonstrated best practices. Sweden organized a study visit in September 2003 to showcase their animal welfare program. The Commission also engaged stakeholders directly, hosting meetings in December 2004 with NGOs and industry representatives, aiming to understand the competing concerns.[172]

At this point, a fortuitous run of Council presidencies created favourable political momentum, as countries that often had animal protection rules exceeding EU norms took the position one after another. Beginning with Luxembourg (January–June 2005), the rotating presidency passed to the UK (July–December 2005), Austria (January–June 2006), Finland (July–December 2006), and Germany (January–June 2007).[173] However, EU enlargement from 2004 to 2007 to include Central and Eastern European countries simultaneously expanded the coalition of governments that had not yet taken major strides in the farmed animal protection space.

The Commission's May 2005 proposal on broiler chickens set maximum density at 30 kg/m², with exceptional circumstances allowing 38 kg/m² if cumulative daily mortality rates met specific thresholds.[174] In the simplest terms, those thresholds meant that if producers limited bird deaths before slaughter, they could keep more of them in the same space. The proposal explicitly referenced both the 2000 scientific report and 'diverging national requirements for the protection of chickens ... [which] has the potential to distort conditions of competition and may interfere with the smooth running of the market organization.'[175] The Commission emphasized that 'legislation came as a response to the long-standing appeal of member states and citizens for the Commission to take action in this area,' while noting that voluntary farm assurance schemes had failed to incorporate key determinants of animal welfare.[176]

Despite the favourable presidencies, resistance emerged from predictable sources. Based on data collected about the member states' positions on the proposal,[177] we can see that a maximum limit of 35 kg/m² was strongly supported by Denmark, Sweden, Germany, and Austria, and that 38 kg/m² was supported by the Commission, the UK, Netherlands, Italy, and Spain. Meanwhile, Belgium, Finland, and Slovenia were in favour of 45 kg/m² and the other member states simply wanted a commitment to return to the issue later, or for no maximum number of chickens per square metre to be set. France consistently sought more socio-economic impact assessments and was joined by newer member countries like the Czech Republic and Slovakia, which prioritized industry competitiveness, while Spain and Portugal questioned consumer willingness to pay for higher welfare standards.[178]

Throughout the rest of 2005 and 2006 there were continued intractable negotiations. Industry stakeholders argued that 42 kg/m² was the lowest limit they could maintain and that they needed a higher mortality threshold[179] – arguing, essentially, that they could only make a profit if they packed bigger birds tightly together and allowed more of them to die early. By February 2006 the European Parliament was torn: the agriculture committee was conceding to 40 kg/m² as an upper limit while the environmental committee was pushing for 25 kg/m² to 30 kg/m². In what might be seen as a small sign of progress, and despite their continued push for higher densities, the agriculture committee was moved to admit that the

> new legislative proposal is a response to increasing public concern about animal welfare. The place occupied by animals in our societies has changed. Despite the industrialization of farming, animals are now seen as sentient beings which have a right to respect.... Public interest in the origin and quality of products has steadily grown with each new epizootic crisis that has arisen.... In the latest Eurobarometer survey (2005), the respondents felt chickens and laying hens to be the two animals most in need of improved rearing conditions.[180]

Some countries argued for delay and more studies, others claimed they were achieving good welfare without regulations, and still others preferred voluntary labelling. Opposition was even expressed by countries that themselves maintained low stocking densities, possibly seeking room for their domestic producers to industrialize in future.[181]

By November 2006, a compromise was reached, supported by fourteen of twenty-seven countries, including major producers like Germany, Italy, the Netherlands, Spain, Sweden, and the UK.[182] However, under the qualified majority rules at the time, these countries lacked sufficient voting weights to force adoption of the new measures, despite representing most of the EU broiler flock. In a last effort to save the proposal, the outgoing Finnish presidency removed any reference to the date at which the stocking densities would be enforced. But even with this change the proposal was defeated at a high-level meeting of member state officials in December, only days before it was due to be ratified by the Council of Ministers. Adding weight to those who stood against the proposal was Romania, which had stated that it would oppose the directive when it joined the EU in January 2007.[183]

The final agreement in May 2007, under the German presidency, reflected significant industry concessions: minimum density of 33 kg/m² with allowances up to 42 kg/m² if certain conditions were met,[184] exactly the level German research had deemed incompatible with animal welfare. Austria, whose presidency had crafted earlier compromises, voted against the modifications on the basis that welfare had been deprioritized and that the 'proposed value of 33 kg/m² for the lower stocking density can … no longer be regarded as a balanced compromise between animal welfare and economic efficiency; it is more reflective of economic interests.' It added that a 'limit of 30 kg/m², as proposed by the Commission and consistent with Austrian national provisions, would have taken more account of the principle of animal welfare.'[185]

Although the final directive established minimum EU-wide standards where none had existed, the outcome fell well short of scientific recommendations and the Commission's original ambitions, illustrating how political constraints can limit even well-positioned reforms. Even with favourable political conditions, supportive presidencies, scientific evidence, industry recognition of harmonization benefits, and majority support among major producers, the dynamics unleashed by the enlargement of the EU, alongside French-led resistance to reform, forced significant compromises.

A SHAKY START, BUT A START NONETHELESS

These early EU animal welfare directives represent some of the world's first comprehensive attempts to protect animals in industrial farming practices through international law. While the outcomes often fell short of scientific recommendations, and animal protection advocates' hopes, they established crucial precedents that no other major trading bloc had achieved. The EU succeeded in banning barren battery cages for hens, limiting the use of sow stalls for pregnant pigs and crates for veal calves, and setting minimum stocking density standards for broiler chickens. These achievements created the world's most advanced framework for farmed animal protection, and the promise of transforming the lives of hundreds of millions of animals across the bloc.

The political compromises required to achieve these break-throughs – extended transition periods, industry-friendly implementation details, and arbitrary thresholds – were reflections of the inherent challenges of building consensus across diverse national interests and economic systems. Nonetheless, these imperfect victories advanced the argument that animal welfare could justify market harmonization and regulatory intervention. They demonstrated the EU's unique capacity to balance economic integration with ethical progress, creating space for member states to lead while gradually pulling others forward through legal obligation and competitive pressure.

Contrary to industry warnings about economic catastrophe, many early-adopting countries discovered that welfare improvements could enhance rather than undermine competitive advantage. The Netherlands provides the most striking example: while French and Italian veal production declined amid consumer resistance to crated systems, Dutch producers who embraced group housing saw their industry flourish. By 2001, the Netherlands exported 95% of its veal to other EU countries,[186] supplying nearly half of German consumption and significant portions of Italian and French markets, precisely because they had responded to consumer welfare concerns that competitors ignored. Similarly, Dutch egg producers maintained strong export positions despite higher welfare standards, while Swedish innovations in enriched cage design became technology exports as other countries sought welfare-compliant alternatives. Even retailers benefited from positioning themselves ahead of regulatory curves: when UK supermarket Safeway promoted high-welfare British veal, sales increased by 30% following publicity about crate cruelty.[187]

These economic success stories demonstrated that improvements in welfare, rather than imposing economic costs, often opened new market opportunities for producers willing to meet evolving consumer expectations. This was a lesson that would prove crucial as welfare-conscious consumers increasingly drove purchasing decisions across European markets.

Welfare directives also revealed the EU's capacity to transcend narrow economic integration and become a vehicle for advancing shared ethical principles that individual nations could never achieve alone. While the single market's primary purpose was eliminating trade barriers and harmonizing commercial standards, the welfare cases demonstrated that European integration could prevent a destructive race to the bottom on social issues. When Denmark or the UK acted unilaterally on battery cages, they faced the classic dilemma of ethical leadership in a globalized economy: domestic producers suffered competitive disadvantages while imports from lower-standard countries undermined both welfare objectives and fair competition. The EU's institutional framework provided a unique solution, transforming nations' moral aspirations into continent-wide legal obligations that protected both animals and ethical producers.

By establishing minimum welfare standards across all member states, these directives showed that European integration could elevate, rather than erode, social standards and create space for collective progress that respected national sovereignty while preventing regulatory arbitrage. This represented a profound evolution in the European project: from a common market designed to prevent war and boost prosperity, to a values-based union capable of codifying shared ethical commitments into binding international law, proving that deeper integration could serve moral as well as economic purposes.

Lastly, the directives were a written commitment that the EU intended to lead on farmed animal protection. The question then became whether those commitments would be more than just words on paper.

CHAPTER 4

MAKING PROGRESS STICK: HOW THE LAWS WERE ENFORCED

The European Union has demonstrated its commitment to the idea of farmed animal welfare and likes to position itself as a global leader in providing such protections. But to truly earn this reputation, the EU must not only excel at creating high standards; it must ensure their effective implementation. This chapter looks at the failures and successes of producers and member states in complying with several major species-specific laws, and the reasons for those failures – reasons that include poorly worded laws, misused transition periods, insufficient inspection and penalty regimes, as well as a lack of market signals like food labelling that allow consumers to reward higher welfare and incentivize compliance.

MEASURING EU NON-COMPLIANCE

Political science literature is replete with assertions that non-compliance with EU legislation is common, citing reasons that range from failures to transpose EU laws into national legislation, to failures to implement laws.[188] Still, despite the evidence provided by existing scholarship, it is actually quite difficult to get a reliable picture of general non-compliance rates in the EU. Many studies use quantitative data on the timely transposition of directives into national law, which does not cover the incorrect implementation or application of laws. Other studies use implementation and application reports prepared by consultancies or academics. But, understandably, due to the resources needed for such work, these studies often only cover a limited number of member states, time periods, and policy sectors.

What we do know is that the European Commission's own numbers show that 45% of pending infringement cases are due to the incorrect application of regulations, decisions, and treaties, while the incorrect application of directives stands at 32%.[189] We also know that agriculture appears to have lower rates of non-compliance compared to environment and taxation, although unfortunately there is no disaggregation of animal welfare policies from general agricultural ones. More specifically, the European Court of Auditors (ECA) has estimated that violations of agricultural compliance standards ranged from 21% to 29% during the years 2011 to 2015, and that half of those violations related to the keeping of animals. The drawback is that this estimate only covered a subset of animal welfare requirements.[190] So instead we are left with doing individual case studies to hint at a broader trend.

EXPLAINING COMPLIANCE WEAKNESS

EU political science and legal literature suggests several theories to explain the bloc's tendency toward non-compliance such that it is, and there are three main schools of thought.[191] First, there are those who argue the main factor in whether a country complies with laws or not is the degree of misfit between the existing legislation in member states and any new legislation proposed by the EU. The second theory looks at whether a 'culture of compliance' exists, based on the legal traditions and administrative capacity of each individual member state.[192] And finally, there is the argument that the number of actors in the system with veto powers – legislatures, regulatory agencies, autonomous regional governments, and so on – is what defines compliance, with delays generally associated with more complex structures.[193]

Each of these theories implies different actions to improve compliance. A 'misfit' perspective might suggest focusing on incremental improvements to narrow the gap between different countries' existing regulations, before setting a floor with EU standards. A 'cultures' perspective offers fewer opportunities for action, because it suggests that compliance with animal welfare standards is unlikely to improve without larger structural changes to historically non-compliant countries. A 'veto player' perspective could, on the other hand, highlight specific actors that need to be targeted, for example, agricultural ministers, farming associations, consumers, retailers, and so on.

Unfortunately, although useful in theory, these ideas are contradictory and based on different metrics, with some measuring how long it took to adopt new standards or whether they were transposed into national law, and others attempting to measure how well these laws were implemented in practice.[194]

The reasons for individual farmer non-compliance appear to be a bit easier to pin down. According to a 2019 survey of the EU's Working Party of Chief Veterinary Officers,[195] a national forum that advises on animal health and welfare, unclear regulations and farmer attitudes are the most relevant reasons for non-compliance in animal welfare. The next obstacles to compliance are insufficient knowledge, financial constraints, and a 'lack of control resources,' a phrase that refers to inspections.[196] These problems abound when we take a more detailed look at how the various species-specific directives have been implemented on the ground.

POORLY WORDED LAWS

The 2019 veterinary officer survey identified unclear regulations as a primary barrier to farmer compliance.[197] To give just one example of a law that embodies this problem, imagine, if you will, that you are an EU pig. As pig farming industrialized, various problems emerged. One of these was that pigs started biting each other's tails, leaving festering wounds. For you, the pig, the wounds are painful. For farmers, they can result in animal infections that are costly and time-consuming to deal with. Because tail biting is the result of multiple factors, including the sex and genetics of the pigs and the way they are kept and fed, it is seen as a problem that is difficult to solve without making substantial changes to the way pigs are raised. A blunter solution involves removing the tails altogether, a practice known as tail docking, which can legally be done without anaesthetics before you reach seven days old as a piglet.[198]

Luckily for you, the law prohibits routine tail docking. What a relief, because your tail is, literally, a bundle of nerves, and having it cut off causes peripheral nerve injury that might be associated with lasting chronic pain,[199] or more plainly: it hurts a lot. Keeping your tail is great for other things too, like balance, flicking things away, and indicating your emotional state.[200] A tucked-in tail might mean you are in pain, afraid, or sick. A curled tail, or a relaxed wagging one, means you're feeling good. Sadly, despite your tail's importance, and the law, studies show that the painful process of tail docking abounds in the EU. Based on numerous sources from the 2010s, it appears that as many as 90% of all pigs farmed in the EU are tail docked,[201] and while there are differences between countries, the average share of docked pigs in any EU nation was estimated to be 77%.[202] The only hope now is that you are a pig in Finland, Sweden, or Lithuania. Why? Because in those countries fewer than 5% of pigs have docked tails.[203]

So what's the problem? If those countries can avoid tail docking, why can't the rest of the EU? The short answer is that loose wording allows for ambiguity, meaning, basically, those who choose to tail dock can do so. Pigs in Finland, Sweden, or Lithuania benefit from the absence of docking traditions in these countries and strict bans on the practice, except for veterinary reasons. As it stands, the EU law only prohibits routine tail docking and allows it if other measures, like providing more space per pig and enough enrichment materials for them to chew on instead – such as 'straw, hay, wood, sawdust, mushroom compost, peat or a mixture of such'[204] – have been tried and failed. Everything about the law leaves room for ambiguity. Was more space tried? Did it fail? Were enrichment materials provided? Did they not help? What does 'enough' enrichment mean? What does 'routine' mean? And it gets more complicated. In the 2010s, the number of pigs with *insufficient* enrichment materials was only 35%, suggesting most pigs are given alternatives to their peers' tails to bite on but still have their own tails docked in most EU countries.[205] So either tails are being docked routinely even though it's not necessary, or the industrial farming system creates conditions so bad that even with enrichment materials it's in the producers' interests to dock pigs' tails.

As the questions and uncertainties pile up, it is somewhat predictable that busy farmers, surviving on thin margins, turn to a one-time, surefire solution: tail removal. Et voilà, your tail is gone, often without anaesthesia, leaving a painful stump that neither wags, nor flicks, nor shows your mood. Sorry, pigs, and sorry, EU citizens, because the failure to actually stop tail docking, despite the existence of a law, is essentially a betrayal of expectations.

Unfortunately, this is just one example of many. The EU's overarching farmed animals directive is so general,[206] the Commission's own fitness check suggested it fails to guarantee even the most basic needs, saying: 'there is still a sub-optimal level of welfare of animals in the EU … [and an analysis] of the legislation and its application shows that this is partly due to the vagueness of certain provisions.[207] For fish the situation is even worse, as we will see in Chapter 7, with laws written in a way that completely ignores the needs of aquatic animals.

TRANSITION PERIODS AS BARRIERS

Transition periods, which were designed to ease implementation of new laws, can instead contribute to delaying compliance and the dilution of legal standards. This is because rather than using transition periods to prepare, producers, and even national governments, often continue business as usual while lobbying for deadline extensions.

This was particularly evident during the run-up to the EU's ban on barren battery cages for egg-laying hens.[208] As we discussed in Chapter 3, in 1999, EU countries agreed that no new conventional battery cages should be installed after 2003, and to decommission existing conventional battery cages by 2012.[209] Thereafter, at least on paper, hens would have to be either cage free or housed in enriched cages. On the ground, however, distortions in the single market plagued the implementation. Austria, Germany, the Netherlands, and Sweden instituted earlier national bans and were already in compliance with the EU law ahead of schedule. While early adoption of higher welfare laws is always welcome, the problem was that countries transitioning earlier were able to sell their conventional battery cage hardware to producers waiting to transition later (see the example of Sweden and Finland cited in Chapter 3).[210]

Another complicating factor was that although the Commission was supposed to give definitive details of what housing was still allowed, along with an implementation date, by January 2005, this did not come until January 2008.[211] A UK Farm Animal Welfare Council report specifically noted that the delay 'has meant that the industry has not felt sufficiently confident to make the commercial decisions necessary to invest in alternatives to the conventional cage system.'[212] By the end of 2004, five years after the directive was adopted, a report provided to the Commission showed that in almost all EU countries there were virtually no hens in enriched cage systems. Instead, citing Finland as an example, it said, 'producers are delaying the decision to invest [in new cages] and the egg industry states that there is still some uncertainty amongst producers as to whether the ban on traditional cages will in fact take place in 2012. Most producers are therefore adopting a wait and see approach.' [213]

In short, many were hoping the egg industry and respective national governments would successfully lobby to delay implementation, and many governments did try that.[214] The delaying attempts failed, however, due to a combination of pressures from leaders in already compliant member states and animal advocacy NGOs, plus a resolution from the European Parliament.[215] The pressure resulted in the Commission reiterating that it would keep to the original deadline,[216] although it appeared to be in a state of denial about the problem, with the commissioner in charge of the ban, John Dalli, reportedly saying '[w]e are not prepared to contemplate that people will not have converted. We think they all will.'[217]

In reality, despite a long transition period, half of EU countries missed the deadline, meaning that between 46 million and 84 million hens spent many extra months or years in tiny cages.[218] If the hens were not kept in the illegal cages, they were culled, with farmers killing up to 33 million birds to meet a deadline they hadn't prepared for.[219] Culls before a cage or farm system change are usually implemented because farmers take the opportunity to make space for younger, more productive birds.

Another example of transition-period failure unfolded around the EU's partial restrictions on extreme confinement of sows. Despite a ten-year transition period, twenty-two out of twenty-seven countries missed the 2013 deadline. Because of this, two to five million of the EU's 13 million sows remained in illegal individual stalls, including 52% to 67% of sows in Germany and France, two of the largest pig meat producers in the EU.[220]

It is hardly surprising, then, that transition periods have been acknowledged by the Commission to be ineffective, at least in isolation, with a European Commission working paper noting that '[t]ransitional periods have not been proved to be very successful.'[221]

It's worth noting, however, that many farmers have been able to make transitions much more quickly than the EU typically requires. For the barren battery cage ban and sow stall restrictions, when enforcement mechanisms were deployed after deadlines were missed, in most cases farmers very quickly complied. In other examples, farmers in Austria and Germany were able to transition away from battery cages by 2009 to meet their national governments' earlier deadlines, and many farmers across the EU have been able to transition quickly to entirely cage-free systems in less than ten years to meet the demands of major retailers.[222]

While one can't expect production methods to shift overnight, much of the historical record suggests the EU does everyone a disfavour by allowing transition times of a decade or more.

LIGHT INSPECTION REGIMES

Inadequate inspection and enforcement mechanisms are another factor that allow legal violations to persist with minimal consequence. The Commission has extremely limited enforcement resources and powers, with farm inspection capacity covering just 1% of holdings annually.[223] In the case of dairy farms, at that rate it would take one hundred years to visit all existing farms in the EU.[224]

Limited EU enforcement resources mean national governments carry the primary responsibility for building systems to prevent, detect, and correct non-compliance, carrying out on-the-spot checks, applying penalties, and recovering funds from farmers who have breached the rules. But national enforcement varies widely, with some countries rarely checking farms against risk indicators, despite this being an EU legal requirement. And, contrary to what might be expected, the variations do not fall along economic lines, with poor inspection regimes limited to poorer countries. For example, to enforce the restrictions on sow confinement in 2012, Dutch and Spanish authorities deployed insufficient inspections, while Polish authorities, by contrast, conducted almost 2,447 inspections and achieved close to 100% compliance some months after the deadline.[225]

Missing data is another problem, as is the difficulty of monitoring the extra conditions that allow producer leeway. A Commission report on broilers found that, among member states accounting for 80% of annual EU chicken meat production, most 'did not have effective and complete systems to monitor, collect and assess information regarding on-farm welfare at the slaughterhouse level.'[226] Another Commission evaluation found that 66% of chickens raised for meat are stocked at high densities.[227] But many countries fail to collect the data that proves farmers are providing the additional welfare conditions stipulated for these higher densities, with inspection authorities often using incorrect thresholds, failing to check gas concentrations, or failing to use the correct measurement equipment.

LEGAL ACTION EFFECTIVE BUT SLOW

Legal action against countries is painfully slow, enabling extended periods of non-compliance. But when it's taken, it's effective. For example, when it became clear most countries had missed the deadline for sow stall restrictions, the Commission demanded they provide regular data updates and asked national ministers to apply sanctions to non-compliant producers.[228] In the space of several weeks, from December 2012 to February 2013, rates of non-compliance fell dramatically: in Belgium from 55% to 10%, in Cyprus and Germany from 52% to 25%, in France from 67% to 25%, and in Ireland from 43% to 10%.[229] Almost overnight, millions of pigs were freed from crates.

On the downside, the infringement process is extremely time-consuming, with cases lasting an average of 45.8 months.[230] If the EU Court of Justice finds a breach, average compliance time, post judgement, is about sixty-one months.[231] This lengthy process allowed 20 million hens in Italy and Greece to remain in conventional battery cages for nearly four years after the official ban,[232] and for several countries to continue using illegal sow stalls for years after the deadline. These included Belgium, Cyprus, Greece, and France, which had to be sent final warning letters.[233]

PENALTIES FALTER

The 2019 veterinary officer survey showed that national inspections, national audits, and EU audits were among the most effective enforcement tools.[234] But the system falters when it comes to fines. As an example, a 2016 report by the ECA found national penalties for non-compliance with requirements are often disproportionate to the seriousness of the violations. It found, too, that the application of penalties for similar cases varied significantly between countries.[235] Meanwhile, the EU's enforcement of the battery cage ban revealed it was worthwhile for farmers in Italy and Poland to pay the fine because the profit from selling the illegal eggs was so large.[236]

In other cases, penalties don't even occur. Indeed, almost all the 2017 to 2019 country reports by the Commission on tail docking indicate that breaches are simply not pursued by national authorities.[237] Reports further suggest that some countries have unwritten agreements not to issue sanctions to farmers or are, at the very least, reluctant to impose large fines that would be contested in court and cause more headaches for national governments.[238] On the upside, when countries provide adequate resources and direction to national agencies, they issue effective penalties more often and more consistently, improving overall compliance.

FINANCIAL BARRIERS AND POOR LABELLING

Next up on the barriers-to-compliance list is money. Funds are an issue both because farmers might not be able to afford compliance (or might not want to spend their limited cash on it), and because there are often insufficient financial incentives for them to do so. Generally speaking, animal welfare is not a very important consideration in farmers' production decisions. This is because – as long as it does not affect productivity – there are no financial costs to lower animal welfare. Improving welfare, on the other hand, usually requires spending money without any guarantee that costs will be balanced by economic gains.[239]

The investment situation is further complicated by the lack of legal labelling standards (with the exception of eggs) that would tell consumers how the animals used to produce their meat, dairy, and fish were treated. The result is that farmers who do invest in higher welfare cannot distinguish their products on shop shelves from similarly labelled products from lower-welfare farms. Absent any information about the conditions in which animals are raised, the lower-welfare item will almost always be cheaper and therefore more attractive to value-conscious consumers.

Poor labelling has another negative effect: it undermines citizens' rights to choose better welfare. Six in ten Europeans say they look for labels identifying products sourced from welfare-friendly farming systems when buying groceries.[240] The lack of labelling also makes it harder for citizens to complain to policymakers about poor welfare conditions. No wonder, then, that the 2019 survey of veterinary officers found 83% favoured an animal-welfare labelling scheme.[241]

CIVIL SOCIETY FILLS THE GAPS

Given the EU's glaring compliance problems, it's no surprise that civil society has often had to pick up the slack and expose what the lack of compliance means for animals. One of the most successful routes to uncovering non-compliance is a complaint system developed by the Commission and national authorities. A 2003 Commission report found the system had developed into the 'chief source for detecting infringements.'[242] The other benefit of the complaint system is that although NGO data provided by this means is not considered official data, and therefore cannot provoke an audit, it may unofficially do just that.

In one possible example of this, the Commission contacted Spanish authorities after CIWF visits found problems on pig farms in Spain.[243] In another, a guilty court verdict for Greece may have been instigated following investigations by CIWF and other animal advocacy groups that showed widespread slaughter regulation breaches.[244] In a third example, the European Parliament debated and published a report on pig-tail docking on the basis of petitions filed by Dyrenes Beskyttelse, a Danish NGO, and Humane Society International.[245] There are some grounds to argue that because most producers eventually comply with a law, it's not useful for NGOs to spend time and money campaigning for compliance. However, these cases suggest NGO pressure has prevented many producers from continuing to operate illegally and harming millions of animals.

RECOMMENDATIONS FOR TURNING WORDS INTO ACTION

Now, instead of a pig, imagine you're a policymaker and you happen to ask my advice on solving non-compliance problems. I'd say two things right up front: 1) design better laws that are both ambitious and enforceable, and 2) invest in the enforcement systems that bring these laws to life. In most cases it boils down to national inspections that prioritize the farms most at risk of non-compliance, combined with fines significant enough to make rulebreakers comply rather than risk having to pay them. If you, the policymaker, wanted more, I might swiftly guide us to the nearest coffee shop, grab a napkin, and make the following list:

Clarity. Laws must be clear, specific, and measurable. Requirements should lend themselves to simple, objective inspection. Think 'no cages' instead of 'monitor floor space area utilized per day to account for species-specific parameters as detailed in annex X of article Y of implementing regulation Z.'

Glidepaths, not waiting periods. Rather than offering decade-long transition periods with regulatory silences that create ambiguity, include phase-out plans with checkpoints where additional support can be offered. One example would be an interim national goal of 20% cage free after two years, 50% after five years, and so on. This would help ensure a managed transition that avoids economic losses and millions of culled animals.

Enforcement. Legislation should include enforcement rules that are translated into national law. These would offer clear penalty guidance, inspection protocols, and reporting duties, and they should avoid complexity by deferring to other regulations.

Strategic funding. Make welfare-related subsidies available for documented improvements, monitor how funds are used, and make them easy for farmers to access.

NGOs and citizens. Encourage whistleblowing and civil-society monitoring. Make complaint mechanisms easier and more transparent.

If the EU got enforcement right, it might avoid a lot of hassle from civil society groups that have been consistently successful at highlighting enforcement gaps through investigations, campaigns, and strategic litigation. The EU can be a leader, not just in protections on paper, but in reality. Thanks to EU policies, there are tens of millions of laying hens, veal calves, and pregnant sows with more space to express their natural behaviour. The system eventually ensured this happened, but it was clunky and delayed.

All we need are some improvements to existing policies, along with a few updated laws for lesser protected species, plus more timely delivery mechanisms, all of which will spare animals, farmers, and EU citizens from the costs – moral and financial – of widespread non-compliance and delayed justice.

CHAPTER 5

A EUROPEAN LOVE
AFFAIR WITH CAGE FREE

Imagine a lineage of hens, each born into a different welfare era. When the European Economic Community was founded in 1957, uniting six nations – Belgium, France, West Germany, Italy, Luxembourg, and the Netherlands – in a vision of peace and prosperity, hens were mostly raised outdoors. These great-grandmothers of the EU's 2025 flock, born in the years after World War II, scratched at the dirt behind farmhouses in newly rebuilding countries. Their lives were short and hard, but they saw the sky. Their daughters were born into a different Europe: one that was industrializing rapidly with a human population that wanted cheap animal protein. By the 1970s and 1980s, as the European single market took shape, and its starry flag began flying over more capitals, most hens found themselves confined, jammed into wire battery cages, unable to spread their wings, dust-bathe, move freely, or escape. While the dream of a united Europe was expanding around them, their world was shrinking to less than the size of an A4 sheet of paper. As the European Union formally came into existence and grew through the 1990s and 2000s, the original hens' granddaughters witnessed the establishment of a common currency and the first set of common standards for hens. Most were still crammed into cages, but gradually they would have seen faint changes: one less cage mate, a perch, a darkened corner in which to lay eggs. These changes were the concrete result of the EU's first animal welfare laws, laws that began to slowly reshape what was acceptable treatment of farmed animals. The hens didn't know it, but people who ate their eggs had begun to care.

As the global financial crisis unfolded between 2007 and 2009, many hens were still confined, but not all. In Austria and Luxembourg, cages were on track to disappear. The great-granddaughters of the EU's postwar hens could once again run, roost, flap, and forage indoors – and sometimes even outdoors. By 2012, an EU-wide ban on the cruellest cages took effect. Progress was real, if unevenly spread among countries. After 2012, there was a lull in any further EU protections for hens. But, in the background, change was happening at retail and national levels. A quiet momentum was building in supermarkets and national legislatures, and more and more generations of hens found themselves never knowing the confines of a cage. And then, starting in 2020, there was fresh hope for improvements. Over 1.6 million EU citizens demanded an end to all cages and the Commission said yes, it would table a legislative proposal by the end of 2023 that would 'phase out, and finally prohibit, the use of cages for all animals mentioned in the ECI.'[246] Scientific advice followed, consultations were held, alternatives were tested. There was hope that all future generations of hens would never know the cruelty of a cage. It seemed the EU was ready to lead the world again, not only in peace and trade, but in compassion.

Only it wasn't. By 2024, as European elections approached, the cage-free dream had stalled. The baton had been picked up, then dropped. This chapter traces the journey of the EU's hens toward a cage-free future and asks: when will the EU finish what it started?

A CAGE IS A CAGE IS A CAGE

EU law in 2025, as laid out in Directive 1999/74/EC,[247] allows the use of so-called 'enriched cages' for egg-laying hens, plus several cage-free systems: barn, free range, and organic. Enriched or furnished cages essentially provide an improvement in living conditions, compared to older barren battery cages, by legally requiring a bit more space for each hen, as well as certain basic 'furnishings,' such as perches for roosting and a nesting box where hens can satisfy their instinct to lay eggs in peace. That instinct is so strong that Nobel prize winner Konrad Lorenz wrote that 'the greatest torment a battery hen faces is the inability to retreat to cover to lay eggs. For the animal expert, it is a truly heartbreaking sight to see a hen repeatedly attempting to crawl under her cage mates in a vain attempt to find cover.'[248] But even with a nest box and other furnishings, a cage is still a cage, and the damage done goes beyond the freedom taken.

The Welfare Footprint Institute has estimated the time an average laying hen spends in different types of pain. The pain categories are based on the intensity and duration of pain caused by key welfare harms, and the prevalence of that pain. The findings show that cage-free aviaries – essentially barn systems with tiered metal structures inside them[249] – are 'clearly superior to conventional cages and furnished cages,' with improvements starting 'soon after a transition to cage-free environments.' The study estimates that during the laying life of a hen kept in an aviary, instead of a conventional cage, 'an average of at least 275 hours of disabling pain, 2,313 hours of hurtful pain and 4,645 hours of annoying pain are prevented.' For each hen kept in an aviary instead of a furnished cage, '1,410 hours of hurtful pain and 4,065 hours of annoying pain [are] prevented.'[250]

And while phasing out enriched cages would clearly benefit hens, much depends, as ever, on the replacement system. Organic and free-range systems, which usually offer more outdoor access than aviaries, are often eschewed by larger producers for various reasons, including higher management costs and, increasingly, avian flu risks due to contact with wild birds. Therefore, the barn system, which offers multiple levels that allow hens to move vertically and horizontally, and which sometimes provide access to the outdoors or to internal 'winter gardens,'[251] has become the most popular cage-free system with meaningful welfare improvements.

But without a clearly mandated higher-welfare alternative, the risk is an industry switch to the cheapest and lowest welfare option, which, in this case, appears to be combination or limited-access barns. These are essentially rows of cages, stacked on top of each other, with internal doors and partitions that can be opened to allow movement or, when inspectors are not looking, closed in a way that resembles battery cages of old. Although it is unclear what type of barn systems are most used in 2025, the egg industry has promoted combination systems, with one article arguing that this is because they can be 'converted from cage-free to conventional, serving as a sort of insurance policy in case the clamour over cage-free eggs dies down.'[252]

As of July 2025, EU egg statistics show there were over 149 million hens still in enriched cages, or about 38% of the total flock, while about 40% are in barns, 16% are free range, and 7% organic.[253] This is a long way off what had been hoped for, with one 2019 report estimating that free-range hens alone would account for up to 65% of the EU's total laying population by 2025.[254] Having said that, the broader trend has been clear for decades: cages are becoming obsolete.[255] As discussed before, there had been numerous calls for the EU to completely prohibit cage systems, stretching as far back as 1981 when the UK government was seeking to phase them out in just five years. And while those efforts ran aground at the EU level, progress continued beneath the surface.

GERMANY'S CONSTITUTIONAL CHALLENGES

In Germany, cage-ban progress is best described as two steps forward and one step back. As early as 1990, North Rhine-Westphalia's state government initiated a constitutional review of hen-keeping regulations, with lawyer Wolfgang Schindler from the Albert Schweitzer Foundation arguing that battery cages violated Germany's Animal Welfare Act by preventing natural behaviours and causing considerable pain.[256]

The decisive breakthrough came in 1999, when Germany's highest court for constitutional matters ruled that regulations permitting 450 cm^2 battery cages were unconstitutional because they violated the Animal Welfare Act on two counts.[257] The first violation was that the cages prevented hens from performing natural behaviours like scratching and pecking; the second was that this produced considerable pain. Following the court decision Germany legislated a complete battery cage phase-out by January 2007,[258] initially planning to allow enriched cages only until 2011,[259] a rule that was stricter than the EU's indefinite permission for enriched systems. The market responded dramatically: cagefree production jumped from 9% in 1996 to 26% by 2003.[260] The bounce was aided by the introduction of EU mandatory egg labelling in 2004 – which made higher-welfare housing more visible in the supply chain[261] – plus retailer initiatives like supermarket chain ALDI's decision to remove cage eggs from shelves.[262]

However, the growth rate in improved welfare for battery hens slowed down, and in 2006 it came close to a standstill.[263] Only organic eggs continued to see increasing demand. Industry pressure led to government backtracking that extended the battery-cage deadline to 2009,[264] allowed enriched cages to exist until 2020,[265] and larger 'colony cages' to continue indefinitely. Happily for hens, in 2008 the Albert Schweitzer Foundation led an alliance of NGOs that called on all remaining supermarket chains to remove cage eggs from their range.[266] The result for German hens was significant, with the proportion of cage-free hens jumping from 40% to over 80% between 2008 and 2010.[267]

In 2010, a German court found the regulation that introduced colony cages violated the country's Animal Welfare Act because the Animal Welfare Commission was not properly consulted, breaching a constitutional requirement.[268] The court gave the government until March 2012 to come up with a solution.[269] Between

2010 and 2011, with pressure from the Green Party, agricultural ministers in the largest egg-producing states of Lower Saxony and Rhineland-Palatinate worked out a compromise between their call for a twelve-year phase-out (with a 2023 deadline), and the federal government's and industries' call for a twenty-four-year phase-out (with a 2035 deadline).[270] The government agreed to a 2025 phase-out of colony cages (with an extension to 2028 for exceptional circumstances), but delayed the final decision until 2015 and allowed enriched cages to continue until 2020.[271]

While political compromises delayed a complete phase-out until 2025, Germany's constitutional framework had created a powerful mechanism for advancing welfare standards beyond both EU minimums and industry preferences, demonstrating how national legal systems could serve as catalysts for accelerating the transition away from intensive confinement.

AUSTRIA'S GRASSROOTS CAMPAIGN FOR TOTAL PROHIBITION

Austria's path to cage-free production demonstrated how sustained grassroots activism and subnational legislation could achieve comprehensive welfare reforms. Starting in the 1980s, when virtually all Austrian eggs came from caged hens, the animal advocacy group Verein Gegen Tierfabriken (VGT) began targeting retailers through exposure campaigns. Their breakthrough came in 1994 when they used ultraviolet light to reveal that 25% of eggs labelled as free range or barn were actually from battery cages,[272] prompting supermarkets to introduce control schemes and leading Billa, an Austrian supermarket chain, to stop selling battery eggs entirely.

Austria's federal structure enabled progressive states to lead reform. Vienna banned cages from 1994, followed by Tirol (2001), Vorarlberg (2003), Carinthia (2004), and Salzburg (2009).[273] Even before that, in 1996, the Austrian Parliament had officially requested EU-wide battery-cage bans, while the federal government provided financial support for farmers transitioning to alternative systems.[274] The decisive turning of the tide began in 2003 to 2004, with a campaign[275] led by physicist Martin Balluch following a conference that united the country's animal advocacy groups. Activists conducted undercover investigations that revealed 79% of Austria's largest farms violated EU regulations. They also rescued hens from battery cages, accepted criminal penalties, and produced films documenting conditions that received extensive media coverage.[276] Crucially, the activists secured support from both major opposition parties, while focusing attacks on the ruling Conservative Party, during one presidential and two provincial elections.

The campaign's tactical sophistication proved decisive. When agricultural officials arranged a positive press visit to an enriched-cage farm in April 2004, activists had coincidentally investigated the same facility the night before, releasing footage of terrible conditions immediately after the official endorsement. At a Conservative presidential candidate's press conference, a Social Democratic writer, who supported the candidate, unexpectedly asked them to ban cages. The following day the candidate switched to publicly supporting a cage ban.[277] Despite Conservative losses in provincial and presidential elections, the ruling party initially resisted complete bans. But sustained pressure from their junior coalition partner, media coverage, and polling showing 86% public support for prohibition, ultimately forced compromise.[278] In May 2004, Parliament unanimously passed legislation phasing out battery cages by 2009, and enriched cages by 2020,[279] while all major Austrian supermarkets voluntarily agreed to end caged egg sales before then.[280]

A CAGE-FREE CHORUS

Across the continent, many other countries took steps forward and back on releasing hens from cages. There was a growing market for cage-free eggs in the wealthy urban populations of Northwest Europe throughout the 1990s, suggesting the need to seriously consider legislation around cage-free systems.[281] By the 2020s, all Dutch and Austrian, and most German, Danish, and Belgian, supermarkets were already cage free for shell eggs (a 'shell egg' is what you imagine when you think of an egg, as opposed to eggs that are processed into a liquid or powder to be used as ingredients in pastries for example).[282]

Contrary to many authoritative claims that Luxembourg had banned all cages,[283] the country's 2002 decree seems to have only prohibited battery cages after 2007, but would have allowed enriched cages.[284] However, Luxembourg had no caged hens of any type by 2006,[285] implying that the very few producers left simply transitioned to cage free or closed business rather than invest in enriched cages. In Sweden, following the government's retreat from cage bans in the 1990s, animal protection group Djurens Rätt looked to retailers directly. Supermarket Hemköp's 1997 decision to eliminate cage eggs sparked a competitive pressure that gradually spread throughout the sector, with over 60% of retailers for ICA (another supermarket chain) following suit by 2008. And when Hemköp reversed course under new ownership, sustained campaigns quickly restored their cage-free commitment.[286]

The Dutch Minister of Agriculture had been pressing for an EU-wide ban on the battery-cage system in the 1990s,[287] but the country's parliament rejected a proposal to phase them out.[288] By 2004, though, many major supermarket chains had stopped selling cage eggs,[289] and parliamentary support to prohibit enriched cages (while still allowing use of larger colony cages) was gaining support.[290] Implementation was repeatedly derailed, however, and then stalled. Despite this, market forces had already driven cage-free adoption to over 85% by 2019 (the rest being caged-egg production for export markets).[291] In Czechia, advocacy group OBRAZ ran a campaign from 2018 to 2020 that created cage-free retailer commitments covering 60% of the market with targeted political pressure.[292] The intensity of public engagement on the issue contributed to the prime minister meeting with OBRAZ and promising to support a cage ban.[293] In 2020, the government agreed to phase out cages by 2027.[294] The Czech Ministry of Agriculture even submitted a proposal[295] to the Commission for an EU-wide ban on cages for laying hens from 2030.

Progress continued across the Union. Neighbouring Slovakia followed Czechia's legislative breakthrough shortly thereafter. The Slovak Minister of Agriculture announced the intention to phase out cages by 2030 as part of a memorandum signed with the egg-industry association.[296] Almost all foreign retail chains operating in Slovakia had already declared in 2017 that from 2025 onward, they would not sell cage eggs to consumers.[297] Belgium's Wallonia region adopted a 2028 phase-out partly responding to broader European trends,[298] though Flanders remained resistant. In Denmark, an initial attempt to phase out cages by 2022 failed, but it was followed by a 2023 law that did just that.[299] The law allowed for a longer phase-out time of twelve years, meaning the country should be cage free by 2035.[300]

In France, President Emmanuel Macron announced a plan in 2018 to phase out cages for shell eggs by 2022,[301] seeking to fulfil an election campaign pledge. Backpedalling and confusion meant, however, that while new installations must be cage free, producers can use a loophole to 'renovate' their existing cages and maintain them.[302] Despite the loophole, France was 70% cage free in 2024.[303] In Poland, long the most resistant to change (the Polish government led a last-ditch attempt in 2010 to push back the EU battery-cage ban deadline), the country's largest retailer, Biedronka, committed to cage-free policies by 2022,[304] while more than half of Poland's thirty biggest egg producers were reported as saying in a 2020 report that they would invest in alternative production, or planned to do so.[305] Perhaps no surprise, then, that the Polish Chamber of Poultry Meat and Feed Producers forecasts that by 2026 the share of cage eggs in Poland may drop to 58.5%,[306] a figure which roughly maps to 43% of hens being cage-free.

In 2003, the UK, still an EU member, rejected calls from CIWF for a total cage ban, which would have allowed a nine-year transition.[307] Nevertheless, at the time of writing in 2025, another CIWF petition has gathered just over 107,000 signatures,[308] and it seems very plausible the UK could jump ahead of the EU by being the first to ban cages altogether.

THE EU PICKS UP THE BATON, THEN DROPS IT

In May 2020 the EU Commission's plan to reform the agricultural section under its Farm to Fork Strategy explicitly promised a full revision of animal welfare legislation.[309] Then, in September 2020, the Commission put out a tender call for the pilot project, Best Practices for Alternative Egg Production. The tender stated that

> public acceptance of the use of 'enriched' cages for laying hens is decreasing, including in the light of scientific evidence showing that such cages severely restrict the ability of laying hens to engage in many normal behaviors. The pilot project will help egg producers meet market demand by providing practical guidance on how to transition to alternative higher welfare cage-free systems.[310]

It added that the project would enable partner countries with many cage-free systems – like Germany, the Netherlands, France, and Italy – to help countries like Spain, Poland, Portugal, and Belgium set up their own cage-free models.[311] On 30 June 2021, the Commission said it would set out 'plans for a legislative proposal by 2023 to prohibit cages for a number of farmed animals … [and] assess the feasibility of working towards the proposed legislation entering into force from 2027.'[312] This was a direct response to the End the Cage Age petition.

In the same year, the European Commission released its own report, *Protection of the Welfare of Laying Hens at All Stages of Production*.[313] Although it made no definite recommendations, under the heading 'Actions from the Commission', it cited the

Commission's positive response to the End the Cage Age petition, and said it was committed 'to propose legislation by 2023 to phase out, and eventually ban, caged farming in the EU for the species included in the initiative.' Just a few months earlier, in April 2021, the European Parliament held a public hearing on the petition.[314] The hearing involved the petition's organizers and three European commissioners. At the session, both MEPs and the commissioners seemed to signal support for a transition. Later, MEPs voted overwhelmingly in favour of a non-binding commitment that supported a transition to cage-free farming.[315]

When the European Commission is considering legal changes to improve animal welfare, it usually first asks the European Food and Safety Authority (EFSA) to produce a scientific report on the issue at hand. For egg-laying hens, that request was duly made and EFSA released its report in February 2023.[316] One of its main recommendations was that birds should be housed in 'non-cage systems with easily accessible, elevated platforms … dry and friable litter and access to a covered veranda.'

In short, by summer 2023 it appeared all institutions were aligned. The science supported a transition, the elected representatives of EU citizens were in support, EU citizens themselves had directly called for a cage-free future, many member states had already begun the transition themselves and were calling on the EU to follow suit. The market was transitioning and the Commission was doing all the necessary preparatory work to bring EU leadership back to the animal protection space.

THE BETRAYAL

Then, in mid- to late-2023, the trajectory shifted. Powerful farming corporation lobbyists, like the European Livestock Voice campaign,[317] supported by COPA-COGECA, Europe's largest farm lobby, mounted an aggressive offensive against the proposed cage ban. Media reports revealed 'ferocious pushback' by industry, including attempts to cast doubt on EFSA and rewrite proposals.[318] In correspondence later obtained using Freedom of Information requests, campaigners revealed that in at least six meetings, between late 2021 and early 2023, the livestock sector repeatedly urged officials in DG SANTE – the Commission's Directorate-General for Health and Food Safety, tasked with drafting animal welfare laws – to 'reconsider ending caged farming.'[319] In the spring of 2023, farmer protests across multiple member states, while ostensibly about bureaucratic burdens and input costs, provided convenient political cover for the Commission to begin walking back its commitments. By summer 2023, DG AGRI, the Directorate-General for Agriculture and Rural Development, was quietly conducting 'stakeholder consultations' that gave disproportionate weight to industry concerns about implementation costs and timelines,[320] effectively allowing opponents to reframe the debate around economic feasibility, rather than the need to keep EU laws in line with science and citizen expectations.[321]

By the end of 2023 the scheduled legislative proposal had not materialized. The results from the EU's official public opinion poll, Eurobarometer, found 89% of citizens opposed cages,[322] but this was released only after it became clear the proposal would not be presented. In November 2023, the Commission quietly dropped any reference to a cage ban from its agenda, instead focusing its animal welfare package on transport rules and pets.[323] The Commission publicly cited the 'highly technical' nature of the issue and the need for more stakeholder dialogue as reasons for the delay.[324] As the weeks passed, commissioners and NGOs began pointing out that time was running short to pass the reforms under the EU's legislative term,[325] due to end in 2024. The elections came and went, with the initiation of a new European Commission at the end of 2024, and without any mention of concrete progress on cages in the work plan for the next commission.

Later, claims emerged that the Directorate-General for Trade (DG TRADE), had weighed in against the welfare reform package for fear it might harm the pending MERCOSUR trade deal with South

America. The fears were, essentially, that South American producers would sour on the deal if they had to ensure their animal-product exports to the EU met even higher welfare standards.[326] The EU-MERCOSUR deal was eventually inked in December 2024.[327]

PROGRESS DERAILED, VULNERABILITIES EXPOSED

The derailment of EU animal welfare reforms represents a critical failure in the Union's capacity to lead on the global stage and deliver on its foundational promise of being guided by values rather than narrow economic interests. But the sophisticated lobbying campaign orchestrated by COPA-COGECA, European Livestock Voice, and allied agribusiness interests didn't just delay legislation; it exposed fundamental vulnerabilities in EU policymaking that threaten the Union's credibility as a regulatory superpower.[328] When industry groups can systematically undermine evidence-based policy through coordinated disinformation campaigns, fund impact studies designed to create fear rather than inform, and strategically exploit crisis moments, it reveals institutional weaknesses that extend far beyond agriculture.

In positioning animal welfare reforms as economically destructive during COVID-19 and the Ukraine crisis, the industry narrative established a dangerous precedent where lobby groups can weaponize external shocks to roll back the EU's policy ambitions. For a union that prides itself on the so-called Brussels Effect – where EU standards become global ones – allowing industry pressure to weaken its standards represents a strategic retreat from global influence.

The failure to advance on animal welfare also undermines the EU's soft-power credentials, at a time when global consumers increasingly demand ethically produced food, while the trade-versus-welfare conflict essentially signals a lack of coherent EU strategy. At its simplest, that lack of strategy means EU trade policy prioritizes trade agreements like MERCOSUR over EU animal welfare, fails to extend any EU protections to farmed animals in other countries, and weakens EU farmers' competitive position in the long term. A truly strategic approach would have recognized that leading on animal welfare standards strengthens rather than weakens the EU's trade position by setting the global agenda.

And yet, progress continues. At the time of writing, Sweden has become 100% cage free, without legislation, thanks to continued advocacy pressure on the market.[329] Slovenia has agreed to phase out cages by 2029,[330] and Estonia will phase them out by 2035.[331] Whether EU leaders like it or not, animal cruelty is being phased out. The question is whether they will continue to stand in the way of progress, or shepherd it more quickly into the present.

CHAPTER 6

A SECOND CHANCE AT PROGRESS

At the time of publication, the current 2024 to 2029 European Union administration bears little resemblance to the pioneering EU of the 1990s that adopted a range of species-specific farmed animal protection policies. Even more importantly, it could be argued, the current administration is significantly different to the previous one that ran from 2019 to 2024 and which offered the promise of a return to progress.

The first major difference is that there is no leadership advancing protections for farmed animals. When the EU lost the UK, a historic champion of animal protection policies, it looked as though it would be replaced with a coalition of member states, MEPs, and commissioners who would make sure Brexit did not mean welfare slipped behind. However, the actions of those who sought and won EU power in 2024 did little to carry this torch. The second difference is that a slow-growth economy is magnifying the ripples of the 2004 enlargement, when the bloc absorbed a range of countries more focused on minimizing the costs of EU integration than animal welfare. While these countries have made many major strides for farmed animal protection since they joined – notably cage and fur farm bans across Central and Eastern Europe – it is these same countries that are threatened most by the economic and security implications of Russia's war against Ukraine. With these concerns heightened, it is much harder for the invisible suffering of farmed animals to break through the policy agenda.

The third difference is that a rightward shift in politics has become unnecessarily intertwined with positions against farmed animal protection.[332] The 2024 European Parliament elections, and the appointment of a new cabinet of commissioners under Ursula von der Leyen's second presidency, brought together a constellation of mostly right-wing political parties, helped into power by an agrarian rebellion. The priorities for this administration were quickly revealed as continuing regulatory rollbacks and putting farmers at the centre of policy debates. (Of course, critics might argue it is the 'big-ag' industry, not farmers, that is being catered to, but that's another book.)

The loss of power among Europe's more liberal and left-wing parties has had a profound impact on animal welfare initiatives. The Greens, classic animal welfare allies, have ceded control to the European People's Party (EPP), a grouping that has branded itself the farmer's party and now holds 'king and kingmaker' positions in the European Parliament.[333] Similarly, in the European Commission, sixteen out of the twenty-seven commissioners are either members of, or affiliated with, the EPP political family, the same family von der Leyen hails from.

All of this has contributed to farmed animal protection falling down the agenda. In the core political documents that set the work programme for the 2024 to 2029 term, the issue of animal welfare barely registers. For example, in the Commission's Political Guidelines there is no mention of animals or anything relating to their welfare.[334] In the Mission Letter to DG SANTE, the word welfare is in the title ('Commissioner for Health and Animal Welfare'), but not in his team's priorities.[335] The only specific welfare direction the letter provides to Commissioner Olivér Várhelyi is both broad and vague, entrusting him to build 'upon the existing animal welfare legislation … [by modernizing] the rules on animal welfare ... while addressing sustainability, ethical, scientific and economic considerations, and citizens expectations.'

More starkly, the Mission Letter for DG AGRI,[336] the agricultural directorate, does not even mention the word 'animal.' And it's not just the Commission that's abdicating leadership. There is no reference to animal welfare in the European Council's Strategic Agenda;[337] moreover, in the Council conclusions animal welfare reforms are rarely mentioned.[338] When they are, the topic is couched in incremental and modest language.

At the time of writing, proposed revisions to farmed animal transport regulations, one of the only 2023 reform package sections to survive so far, have been slow-walked by both member states and MEPs. The transport reform risks one of two unappealing options: the revisions being dropped entirely, or the regulation being worsened for animals by the inclusion of different revisions.[339]

ON THE (SLIGHTLY) BRIGHTER SIDE

Despite the apparent paucity of salient support for farmed animal protection in the 2024 EU administration, signs of life can be found across the political spectrum, at national level, and within the layers of EU governance.

A total of 115 MEPs who signed the Vote for Animals pledge were elected in 2024.[340] Seven EPP MEPs belong to the Parliament's Intergroup on Animal Welfare, an informal gathering that promotes better protection laws. And two of those seven are additionally vice-chairs of the Parliament's agricultural committee, meaning they have a voice in EU agricultural policy planning. This indicates that even within the right-wing EPP there is some support for welfare reform. Ten different political parties formed specifically around animal protection fielded candidates in the 2024 European Parliament elections, attracting a total of 1.5 million votes,[341] and two were elected,[342] signalling a small but vocal group of citizens who are driven by this issue above all else. As more citizens realize they can contact their representatives and tell them they care about the issue and expect progress on it, the more incentive there is for champions to emerge.

Indeed, at citizen level there continues to be a great deal of public support for farmed animal protection. In March 2023, a Eurobarometer survey found that 84% of Europeans believed the welfare of farmed animals should be better protected than it is now, and 83% supported limiting animal transport times.[343] In November of the same year, another survey – this time by the European Consumer Organisation – found about nine in ten of those surveyed considered it was important to implement 'new laws to improve the welfare of farmed animals such as providing more living space, banning cage systems, and mutilations.'[344] Another Eurobarometer, published in January 2025, showed that almost nine out of ten people surveyed agreed that 'agricultural imports of any origin should only enter the EU if their production complies with the EU's environmental and animal welfare standards.'[345]

Beyond surveys – useful, but not always reliable, indicators of public commitment – actions are being taken by national governments, indicating there would be support if the Commission chose to act in this arena again. Recent national initiatives include a 2024 move by Belgium to recognize animals as sentient beings,[346] a 2021 German ban on the culling of male chicks in the egg industry,[347] and, in 2018, a ban in Luxembourg on killing animals for economic purposes, such as male calves from dairy cows.[348]

QUESTION FOR A NEW PARADIGM

Overarching both the broader political challenges and the smaller signs of progress is one key question: Of all the potential – much-needed and overdue – improvements to animal lives, which initiatives are both the most pressing and the most feasible? Or, to put it another way: what are the policy initiatives that will save this administration, and possibly the next, from completely dropping the ball and instead ensuring they make progress toward aligning with citizen's expectations, market trends, national laws, and the latest animal welfare science?

Part of answering this question involves acknowledging that the EU's policy direction has been trending toward a de-bureaucratization that is at risk of becoming deregulation, and away from producing any new regulations. In one example of this trend, the Vision for Agriculture and Food, released in February 2025, suggests the Common Agricultural Policy will focus more on incentives than regulations, or 'impositions' as Commissioner Christophe Hansen is reported to have described them.[349] In another example, President von der Leyen said in her political guidelines for the new Commission that '[w]e must enable farmers to work their land without excessive bureaucracy.'[350] At first glance, this might not seem the most conducive climate to greater farmed animal protections.

Difficult as the current political paradigm might be, it does not have to be a death knell for animal welfare policy improvements. The two most important commissioners when it comes to welfare, Várhelyi and Hansen, have signposted interest in gradual, species- and sector-specific change when economic concerns are neutralized. For example, Várhelyi said during European Parliament questions that '[o]ur overall approach [to the End the Cage Age ECI] will be balanced and gradual [and] will include accompanying measures to support stakeholders in making this transition.'[351] For his part, Hansen, has committed to improving farmed animal welfare, which he said has 'tangible benefits for farmers, notably the reduction of risks of outbreaks of diseases, as well as the reduction of the use of medicinal products.' To do that, he further committed to working 'closely with the Commissioner for Health and Animal Welfare, notably to modernise [welfare] rules […] in line with scientific, environmental, economic and social factors.'[352]

This means that there is space for farmed animal protection, even in a political climate geared toward protecting farmers from burdensome regulations. The question now is: which protections and policies already have momentum behind them and make perfect sense in the new era of competitiveness?

WELFARE AS A WIN NOT A BURDEN

During the 2019 to 2024 EU administration, animal welfare was hitched to the wagon of the Green Deal, an ambitious programme to set the EU on course to meeting its targets to curb climate change. The drive to continue improving conditions for farmed animals was included in the mission of the Green Deal's Farm to Fork strategy to make the food supply chain more sustainable.[353]

This was a benefit to animal protections when the Green Deal was popular. But the animal protection proposals were placed among the last the Commission would deal with during its time in office, and by the time they were addressed in 2023, both the Green Deal as a whole and the Farm to Fork strategy in particular were sailing in choppy waters. Farmer protests were used as a launchpad to collapse multiple EU environmental policy proposals in 2023 and 2024, and apparent concerns over food security were used to justify killing the Commission's pesticide reduction law and exclude most factory farms from the industrial emissions directive.[354] By February 2024, health, food, agriculture, and politics journalist Gerardo Fortuna estimated that '63% of the ongoing initiatives of the EU's flagship food policy [were] basically dead in the water – namely not proposed yet or even withdrawn.'[355] A subsequent humorous post from Fortuna declared the time of death of the Farm to Fork strategy as 12:43, 19 February 2025.[356]

Animals' hopes can no longer rely on environmental protection promises. Environmental policies that are seen as impinging on farmers have become so sensitive they risk damaging progress for animals. Furthermore, the macroeconomic conditions created by inflation, Russia's war in Ukraine and generally sluggish economic growth are encouraging the EU's paradigm shift away from top-down regulatory requirements and toward bottom-up deregulation incentive schemes. Thus, the progress from the EU may not be coming from bureaucrats in Brussels who are detached from the situation on the ground.

All of this means farmed animal protection needs to be recognized as a policy issue in its own right, and one that functions as an economic win for farmers, not a costly burden imposed by policymakers with little consideration for economic realities. These wins would include developing opportunities to support income generation from higher-welfare product sales. One example of this would be clearer labelling schemes, ones that show consumers the kinds of animal welfare conditions they are paying for, along with any opportunities, such as clauses in trade deals, that would protect EU farmers from being undercut by lower-welfare imports. To support the switch to higher welfare conditions, farmers need financial incentives, and those incentives should be easily accessible and organized in a way that avoids any last-minute stumble to the finish line, and certainly not the stumbles that have plagued previous EU species-specific protections. At the same time, in line with studies on financing the cage-free transition,[357] national agriculture ministries and DG AGRI should be pushed to spend the EU's Common Agricultural Policy funds to support transitions and ensure financial concerns are not a reform barrier. After all, if the EU believes optimizing competitiveness is key to unlocking economic growth, it must consider the gains to be made from a compassionate food supply.

MAKING THE CASE FOR HENS

There are signs, as discussed in Chapter 5, that despite animal welfare being a low priority, the EU's legal foundations are moving toward a cage-free future for egg-laying hens.

Under the EU's Better Regulation Framework, the 'one in, one out' principle requires that new regulatory burdens be offset by removing or reducing existing ones.[358] This creates a political challenge for introducing new farmed animal welfare regulations, as policymakers may be reluctant to add rules without clear offsets.

Happily, implementing a uniform cage ban for egg-laying hens would simplify regulations by removing the burden currently associated with regulating two systems, caged and uncaged. And, as a once-off change to fixed infrastructure, cage removal would also have less regulatory and enforcement complexity than ongoing compliance-based reforms that involve daily monitoring of management practices or animal welfare outcomes.

Another factor favouring an EU cage ban is that the political risks normally associated with a welfare reform are offset by the market demand for cage-free eggs. As of July 2025, about 62% of EU egg production was already cage free.[359] Plus, ten EU countries already have cage bans, or are almost entirely cage free.[360] On the process side, there is another plus: previous impact assessments, including the cage-ban proposal, have already passed the Regulatory Scrutiny Board process, reducing the administrative burden on the Commission. In fact, once all the factors above are taken into account, there is really no significant reason not to go ahead with a cage ban. This is an area the EU has already legislated, meaning a ban would be the continuation of that work, and it would build on existing experience and capacity. Furthermore, countries are asking the Commission to take this action, mainly to solve the EU market distortions being created by individual country actions taken in response to consumer, market, and scientific expectations.

There is another attraction, too: a clearer conscience. After all, how many generations of hens trapped in cages do current EU leaders want to have as part of their legacy? Wouldn't they rather be able to say the first generations of entirely cage-free hens in the EU lived under their watch? If so, they need only give an open door a gentle shove. And ideally, based on the experience of past reforms and for the sake of all concerned, they would be sure to avoid any lengthy transition times.

IMPORT HARMONIZATION

There is no good reason for the EU to keep its animal welfare leadership within its borders. When it chooses to act as one entity, the EU is one of the largest economic forces in the world. The problem is that the farmers and consumers who contribute to that economic force may be reasonably concerned that while they work and buy in ways that improve conditions for animals, those efforts risk being undermined by imports of cheaper, lower-welfare products from animals living worse lives outside the EU. More broadly, if retailers can stock shelves with lower-welfare products, all citizens, farmers included, are robbed of their democratic right to an EU that is a force for good.

To counter these risks, the EU should, at the very least, ensure that animals farmed in conditions worse than the EU currently allows are not imported into the single market. Better yet, the EU should encourage external producers to comply with EU standards, a move that would raise welfare globally. Most importantly, the EU should clearly communicate its commitment to protecting farmers against lower-welfare imports. Already, Hansen, as the EU commissioner responsible for the EU's Vision for Agriculture and Food, has made it clear that he wants to 'pursue, in line with international rules, a stronger alignment of production standards applied to imported products, notably on pesticides and animal welfare.'[361] Great! Keep telling us that. And tell us what you are doing to make those words a reality.

The same can be expected of Várhelyi, who said that he would work globally 'to promote high international standards' via multilateral forums like the World Organisation for Animal Health (WOAH) and with countries that are candidates to join the EU. On the trade front, he has suggested adding animal welfare clauses to trade agreements that would ensure 'imports into the EU of live animals comply with EU rules or equivalent standards, thereby fighting the race to the bottom in animal welfare standards globally, which will be contrary to our citizens' expectations.'[362]

He has also made it clear he 'would pay particular attention ... to designing a balanced approach on animal welfare ... [that would] respond to citizens' expectations and which are also economically and financially viable for farmers, and which does not hinder their competitiveness.' [363]

For both commissioners, and the EU in general, any progress on preventing lower-welfare imports would be doubly attractive because it would demonstrate that the EU is a major economic power, one that helps raise standards around the world by demanding imports meet higher welfare requirements.

Even more importantly, perhaps, if the EU were to lead by phasing out cages for egg-laying hens, and ensure those standards are exported abroad, then it would surely regain its stature as a leader in this space. This would not be unprecedented. In July 2023 the EU signed a trade agreement with New Zealand that required all animal-source products to be produced under standards equivalent to EU rules. And in December 2023 the EU signed an interim trade agreement with Chile whereby the two parties agreed to cooperate on implementing animal welfare standards on farms, during transport, and at slaughter.[364] The EU has also banned seal fur imports, in the face of public outrage, and the ban was upheld by the World Trade Organization.[365] Looking ahead, the proposal for a new regulation on dog and cat welfare is aiming to cover imports.[366]

In short, these are common-sense reforms with multi-stakeholder support that would help build momentum and prevent the EU's welfare ambitions petering out, like many times before. And if the EU cannot make such simple high-reward, low-risk changes, then it will be no wonder if citizens, the market, and scientists continue to lose faith in its institutions and look elsewhere for legitimate representation of their expectations.

CHAPTER 7

FARMED FISH WELFARE: THE NEXT FRONTIER FOR EU LEADERSHIP

There's yet another world under the surface. When it comes to the suffering they endure before they land on plates, the world's fish can lay claim to being one of the groups of farmed animals most often forgotten, both by lawmakers and the general public. This means that despite decades of progress that put limits on various forms of extreme confinement for terrestrial animals, farmed fish remain conspicuously left behind in the EU's march toward a more compassionate union. On the other hand, this precedent presents a clear avenue for progress once momentum returns.

FISH COUNT

As things stand in 2025, even the basics of counting fish is a neglected issue, with the EU typically numbering them in tonnage rather than counting them as individual animals. This fundamental misunderstanding was evident in one of the EU's first fish welfare events in 2018, organized by Eurogroup for Animals in Brussels. At this meeting a member of the directorate responsible for animal welfare, DG SANTE, had to be educated about how to calculate the numbers of different kinds of farmed fish: while Atlantic salmon made up the majority of the EU's farmed fish production by tonnage, as individuals, gilthead sea bream and European sea bass are the most numerous.[367] Because each salmon is so large, Europeans get more meat from a single salmon than from a fish of many smaller species. Gilthead seabream and European seabass are much smaller than Atlantic salmon, so producing even a smaller total tonnage of seabream or seabass meat requires killing far more individual fish.

Lacking even a proper estimate of the size of the problem, it's no wonder the EU has not stepped up with effective solutions. While we lack official figures from the EU, the latest high-quality estimate is that between 670 million and 1.1 billion fish were farmed and killed in the EU in 2022.[368] That is more than the combined total of all the pigs, turkeys, ducks, sheep, cows, and goats farmed and killed in the EU every year.[369]

Salmon is the fish most consumed by tonnage and most popularly eaten across the EU.[370] Compared to other farmed species, salmon benefit from the largest number of legal and corporate protections guaranteeing proper stunning before slaughter, as well as a range of other protections for their on-farm welfare. There's just one problem. Most salmon purchased by EU consumers are farmed in Norway and Scotland, both non-EU members, and exported to the single market. Ergo, the farmed fish of preference among Europeans, and the one that receives the highest level of protections *within* the EU, are those that are mostly farmed *outside* the EU. If ever there was an embarrassing gap in EU leadership on farmed animal welfare, this is it. Not only is the EU not protecting their higher-welfare farmers from lower-welfare imports of pig and chicken meat, they're not even managing to compete with higher-welfare fish imports.

In terms of EU production, the four most farmed fish species in the EU by number of individuals are rainbow trout, gilthead sea bream, European sea bass, and common carp.[371] Most EU fish farming happens in just a handful of countries: France, Greece, Spain, Italy, Croatia, and Cyprus, and the type of fish farmed varies by country.[372] While trout farming is concentrated in Northern Europe, and carp farming in Central and Eastern Europe, most sea bass and sea bream are farmed to slaughter weight in Greece and Spain. Most EU consumption of these species is supplied by EU farms, not imports from outside the EU, as is the case with salmon.[373] This means the welfare of these fish is a homegrown problem for the EU, particularly for sea bream and sea bass, with the bloc accounting for about 45% of the world's production,[374] or about 800 million juveniles annually.[375] By the same token, the EU's key role in producing and consuming these fish means it has the power to improve the lives, and deaths, of millions of sentient beings, and to take a leadership position while doing so.

FISH DISMISSED

The relative invisibility of fish in considerations of farmed animal welfare has been attributed to their lack of the human-like characteristics that can evoke empathy for land animals. Their faces are not expressive and, without vocal cords,[376] they cannot scream. But scientific studies suggest fish show clear responses to pain[377] and are capable of suffering.[378] A 2003 study by biologist Lynne Sneddon found that fish have nociceptors, sensory neurons that detect harmful stimuli.[379] A 2019 study by the same researcher showed fish who are given pain relief following injury resumed normal behaviour, indicating they experience true discomfort rather than just unconscious reflexes.[380]

Neurobiologist Culum Brown has further argued that one of the most interesting forms of evidence for marine animal suffering is that pain 'appears to distract fish and prevents them from carrying out other tasks or paying attention to external stimuli.'[381] Most humans can sympathize with that. The European Commission itself acknowledged that 'there is now sufficient scientific evidence indicating that fish are sentient beings and that they are subject to pain and suffering.'[382] Despite this acknowledgement, the EU has done little to protect these animals from that pain and suffering.

NOT DESIGNED WITH FISH IN MIND

EU legislation governing on-farm conditions, transport, and slaughter was primarily designed for terrestrial animals and fails to address the specific needs of fish. There are no species-specific welfare standards for fish on farms, no requirements for water quality, stocking density, or enrichment, leaving billions of fish vulnerable to chronic stress, injury, and poor husbandry.

During transport, fish are covered only by general rules that were not written with aquatic animals in mind. For example, the existing regulation considers animals unfit for transport if 'they are unable to move independently without pain or to walk unassisted.' When was the last time you saw a fish going for a walk, assisted or not? The regulation also makes requirements about floor and head space and air quality. There is nothing about water quality, travel stocking densities, or much of anything else that matters for an aquatic animal. This results in overcrowding, low oxygen levels, and handling practices that often cause serious suffering.[383] At slaughter, the regulatory gap is even starker. Fish are explicitly excluded from the regulation's species-specific slaughter rules. In practice, this means it is legal to kill fish in ways that would be unacceptable for warm-blooded animals. These inhumane methods include death by suffocation (meaning fish are left out of water until they die) or death from being dropped into ice slurry (a freezing mix of ice and water). Scientific bodies like the EFSA have condemned both practices for causing prolonged suffering.[384]

And yet despite the legal and administrative invisibility of fish in the EU system, the public and many stakeholders in the aquaculture supply chain are increasingly aware of, and vocal about, farmed fish welfare. As early as the 2000s, we can see that public concern for fish welfare appears to have been surprisingly high, similar to farmed land animals like pigs and chickens.[385] A 2020 public consultation on EU aquaculture found 67% of respondents felt fish welfare should be improved.[386] When consumers were asked about fish welfare in a 2020 survey,[387] the percentage of respondents who said that clean water and humane slaughter are essential for fish welfare was 95% and 89% respectively, while 61% said fish welfare would influence their buying decisions. A year later, in March 2021, the Commission published its evaluation of the 2012 to 2015 animal welfare strategy, writing: 'Stakeholders from across different disciplines believed that legislation and enforcement related to fish welfare needed to improve. The increased interest in fish welfare among consumers indicates that there is public support for this.'[388]

ETHICS, CREDIBILITY AND OPPORTUNITIES

Given the scale of the fish farming industry, and the EU's keen interest in expanding it,[389] it is ethically essential to address and prevent the cruelties currently inherent in the system. The scale of both fish farming and citizen support for better welfare means the absence of legal protections for fish is both an ethical failure and a credibility risk for the EU. The bloc cannot claim to lead on animal welfare while excluding one of the largest groups of animals in its farm system from meaningful welfare policy. For a union that has led global commitments to higher welfare food systems, ignoring farmed fish is both inconsistent and unsustainable.

Fortunately, this gap is an opportunity. There are several factors that support the adoption of legislation to better protect aquatic animals, providing a clear foundation for the EU to extend its leadership. These factors include upcoming EFSA opinions on fish welfare, due in the late 2020s, as well as growing consumer support, proven models from other countries, and EU-funded pilot projects.

EU OR NATIONAL

As noted earlier, the species farmed differ by country and region in the EU. Because fish farming is not an issue common to all EU countries –in the same way, for example, egg-laying hen and broiler chicken farming is – the question arises whether the EU has competency to act, or whether the issue should be the national responsibility of each relevant member state.

The answer is: it depends. If individual EU countries were making progress toward developing publicly funded fish stunning trials, incorporating humane slaughter into national certification systems, and agreeing welfare guidelines, then there would be a much weaker case for the EU to spend resources and exercise leadership. But that type of national progress is not happening. Instead, as of 2025, the Netherlands, which is not a significant fish farming country, stands out as one of the only member states that has pursued fish welfare legislation incrementally since 2002, when the Dutch Council on Animal Affairs was commissioned to develop assessment frameworks for new farmed species. Key milestones include the 2007 Animal Welfare Memorandum, establishing welfare objectives for aquaculture husbandry, transport, and slaughter; the 2010 Sustainable Aquaculture Benchmark, offering tax benefits for farms meeting welfare standards; and the 2016 Assessment Framework for Production Animals, extending welfare protections to fish.[390]

Germany, Sweden, Ireland, and Spain each have laws or guidelines that in theory protect fish at the time of killing, but there is heterogeneity in what is classified as permissible and what is enforced. These inconsistencies are highlighted by the 2019 Greek industry guidelines. These set welfare expectations and acknowledge that the 'pain and anxiety inflicted' by killing should be kept to a minimum, but then immediately admit that slaughter using ice or ice water – which is 'not considered humane' – is still used.[391] Furthermore, although there are geographic differences in terms of which fish species are farmed and consumed, there are still implications for the single market because much of the fish farmed in the main EU producer states is consumed in other EU countries. One study found, for example, that '92% of sea bass and sea bream produced in Greece is exported, with Italy, Spain, and France being the largest markets,' while about 90% of Denmark's trout production is exported, much of it to Germany, Poland, and Finland.[392]

Once again, this means that if producers in some countries make progress in protecting animals through national legislation, they will risk being undercut by lower-welfare imports from other member states. Similarly, if citizens demand more protections for fish in their own country, but the market is flooded with fish from other EU countries without these protections (and without clear labelling to identify the fish with better living conditions) the single market becomes a vehicle for cruelty, not progress. Nor is this an EU-only problem, because the bloc's farmed fish consumption 'consists of roughly one-third domestic (EU) production and two-thirds imported fish.'[393] Therefore, if the EU chose to apply any standards it adopts for fish farmed within the bloc to imports, this would be an area in which it would protect domestic producers from unfair competition from lower-welfare imports and could stimulate countries around the world into raising their own standards.

In this context, it is worth remembering once again that it is unreasonable to expect consumers to bear the brunt of ensuring the industrial animal farming system adheres to welfare science recommendations, especially when the farming methods are kept hidden from view. The responsibility lies with institutions, and their leaders, to carry out the wishes of the citizens.

PROGRESS SO FAR

European Union efforts to protect farmed fish have evolved through two distinct phases since the 2000s. Initially, as scientific consensus emerged around fish sentience, the EU prioritized boosting its aquaculture sector while conducting foundational welfare research.

In 2004, EFSA issued critical opinions stating that for many farmed species, no commercially acceptable humane slaughter methods existed.[394] This was followed by comprehensive scientific reports in 2008 and 2009 that covered major farmed species like salmon, trout, and sea bass. Early legislative attempts to build on awareness of fish sentience stalled, however, when the 2009 Council Regulation on animal slaughter excluded fish and the Commission deferred action until a mandated 2014 study could be completed.[395]

Despite EU delays, the international governance system continued to move ahead. The Council of Europe adopted fish welfare recommendations in 2005 and the WOAH made a similar move in 2008.[396] The 2010s marked a shift toward more structured EU institutional engagement in fish welfare, with the establishment of key advisory bodies, including the Aquaculture Advisory Council in 2016,[397] a body which brought together industry and NGO stakeholders to provide formal recommendations to the Commission.

In 2017, the Commission's long-delayed report on fish slaughter was finally released.[398] It concluded that voluntary measures and member state-level actions were preferable to EU-wide harmonized legislation, citing the sector's diversity and developing technology. But, by that time, public opinion had progressed and there was growing support for EU-wide standards. Two years later, in 2019, the European Parliament held its first dedicated fish welfare debate and thirty-nine MEPs called for harmonized EU standards and legislative proposals.[399] In December that year, the Council formally requested the Commission to assess new legislation covering all species without existing protections, explicitly including farmed fish.[400] In June 2020, the EU Platform on Animal Welfare published best practice guidelines on water quality and handling for the welfare of farmed fish.[401] The guidelines were developed by a working group led by Greece, together with Spain, Italy, Germany, and Denmark, plus non-EU member Norway. Most recently, the promised but thwarted 2023 farmed animal welfare reform package would likely have included increased slaughter protections for the major farmed fish species. But in 2024, the head of DG SANTE suggested any further protections for fish would have to wait until more EFSA opinions were issued

in the late 2020s.[402] Better to wait, apparently, until the EU is behind not just one decade of scientific opinion, but two.

A WEALTH OF PROTECTIONS TO DRAW ON

If and when the EU does decide to regulate, it will not be starting from scratch. Steps to harmonize the market – by regulating
stocking densities of organically farmed fish, for example – have
already been taken.[403] These efforts prove that the EU can act on
aquaculture where leadership is needed. However, organic production makes up only a few percent of total production.[404] Moreover, when it comes to drafting species-specific laws, the bloc
could easily draw upon a range of existing work. These include the
WOAH policies that were laid out in 2021,[405] the EU Platform on
Animal Welfare water quality and handling guidelines, published
in 2020,[406] and the 2006 Council of Europe Recommendation
concerning farmed fish.[407] If these policies were clearly encoded
into EU law it would help on three fronts. First, it would remedy the fish welfare legislation gap. Second, it would end ambiguity around the applicability of various voluntary, loosely worded
standards, guidelines, and other documents. Third, it would es-
tablish a clear set of legal standards that producers across the EU
could expect competitors to adhere to, discouraging those who
produce cheaper products by compromising welfare.

EU policymakers could further draw on existing species-specific regulations in a range of countries, including slaughter regulations in Norway (Art. 14),[408] Switzerland (Art. 184),[409] Germany
(Annex 1, 9),[410] and New Zealand (Part 6);[411] stocking densities in
Maine (L.D. 1951),[412] Norway (Art. 25),[413] Switzerland (Annex 2,
Table 7),[414] and Chile;[415] water quality standards in Switzerland
(Annex 2, Table 7)[416] and Czechia (Annex 5);[417] and water quality
discharge standards in Turkey (Annex 5).[418] The range of existing
laws in these countries is not only a useful resource for the EU; it
is a clear indication of how far the bloc now lags behind on fish

welfare legislation. Instead of worsening that lag, the EU needs to catch up and, ideally, start setting the pace.

HUMANE SLAUGHTER

Stunning fish before slaughter, a policy already standard in high-welfare terrestrial farming, is supported by leading scientists, NGOs, and certification bodies such as the Aquaculture Stewardship Council and EU Organic. But this technique is not often employed. According to the European Commission's leaked impact assessment, which would have accompanied the 2023 reform package, fewer than 5% of the EU's farmed sea bass and sea bream are humanely knocked unconscious before slaughter to avoid unnecessary stress and suffering.[419] Instead, many fish die a prolonged death in ice slurry, a practice EFSA says should be phased out. For trout, the impact assessment estimates that between 20% and 50% are stunned, but it is possible this is an over-estimate because many trout are small and the percentages are based on volumes rather than individual fish numbers.

Given the urgent need to prevent so many agonizing deaths, it is important to note that a number of successful case studies show the feasibility of transitioning to humane stunning methods in less than 10 years – the timeframe the EU's 2023 reform package had been considering. These case studies include examples from Ace Aquatec, a Scottish company specializing in stunning and culling systems, which has successfully implemented sea bass and sea bream stunning in Greece and Turkey. It also offers options for rainbow trout.[420] Other organizations and guidelines promote better slaughter standards. These include the Aquaculture Stewardship Council,[421] a non-profit certification group, and the EU's organic standards,[422] which stipulate that slaughter methods 'shall render fish immediately unconscious and insensible to pain.'

Added to the welfare benefits, there are studies that show humane stunning can be implemented without significant economic drawbacks, making producer uptake more probable if legislation required it.[423] One point to note, however, is that while welfare-validated commercial stunning equipment exists for salmon, the same cannot yet be said for sea bass and sea bream, potentially delaying the enforceability of any stunning legislation for these fish. But even with delays, the EU can demonstrate leadership by commissioning studies and trials on stunning equipment, a move that would support the development and utilization of world-class welfare technology. In fact, legislation requiring effective stunning would likely incentivise the market to innovate and expedite technological improvements.

REARING

Apart from stunning, there is another area in which the EU could vastly improve fish welfare: daily living conditions, especially water quality, oxygen levels, and stocking densities. These are among the most important welfare determinants and are currently unregulated beyond the organic sector.

That lack of regulation means the EU's farmed fish face a range of problems.[424] The most common of these are overcrowding, and temperature and oxygen ranges that fail to meet the needs of the fish. The former usually occurs because, just as terrestrial farmed animals are often packed into a minimum number of square meters, the more fish housed per cubic metre (m^3) of water, the greater the potential profit. Commercial stocking densities for sea bass and sea bream can reach 20 kg of fish mass per m^3 and up to 30 kg per m^3 for trout. The EU's organic standards, by contrast, stipulate more comfortable densities of up to 15 kg/m^3 for sea bass and sea bream, and 25 kg/m^3 for trout.[425] Legal precedents that could pave the way for change include existing stocking density limits in the EU organic standard and associated labels, such as those in the certification schemes from GAP international Certification and Naturland; laws in Norway and Chile; and the standards set by the UK's RSPCA. In temperature terms – with higher temperatures meaning less oxygen – the 2019 Greek industry guideline for sea bass and sea bream cites EFSA parameters of 2°C to 35°C for sea bass, and between 5°C and 34°C for sea bream.[426] These are wider by about 10°C either way than CIWF recommendations, which suggest the optimal range for both species is between 18°C and 26°C.[427] On the oxygen side, the 2019 Greek industry guidelines cite EFSA, suggesting 'oxygen saturation in rearing conditions should not drop below 40%.'[428] In contrast, CIWF recommends keeping as close to 100% as possible, with 70% to 110% as an acceptable range.[429]

And while temperature and oxygen are difficult to control, the urgency to do so is only rising. The same 2019 guidelines acknowledge that research shows some farms are already approaching oxygen limits in summer and, as with any other area of animal welfare, that situation is likely to be generally aggravated by climate change, and specifically by marine heatwaves.[430]

None of this is straightforward in terms of monitoring. In the simplest terms, enforcement would require mandating record keeping, data inspections, and fines for not correcting any trends that degrade fish living conditions. These systems are not impossible, though they could be seen as onerous by producers.

Items that are easiest to measure include temperature, salinity, and pH, and none call for specialist staff. Policymakers could also draw on the Swiss law (Annex 2, Table 7) that sets dissolved oxygen, oxygen saturation, ammonia, nitrate, saline, carbon dioxide, pH, and water temperature requirements for salmonid and cypriniform fish.[431] Temperature control could additionally be made easier with automated monitoring, although this would cost more. The other problem with temperature is, of course, that even when producers know it's too high or low, it's hard to change it in the sea cages where most of the EU's sea bream and sea bass are farmed. This may have wider implications for where fish are farmed, or not farmed, in the EU. Oxygen is harder to measure and automated systems are costly, meaning this is an area where it is more feasible to set standards for indoor hatcheries and fish farms on land, both of which have greater control of water temperature and therefore oxygen.

A commonly cited barrier to reform of any kind is cost. A 2013 analysis found that fish farmers who meet the EU organic standard spent about 30% more than those who did not.[432] But because the organic standards involve a wider range of requirements, the cost increases for individual welfare benefits might not be as high as that. Another factor that may reduce the additional costs is that most of the equipment is probably already in place, given that producers monitor various aspects of water quality for other reasons (e.g., environmental protection).

With this in mind, the most expensive factor may be hiring additional staff to monitor data trends and mitigate them by, for example, using aerators and cleaning nets, lowering stocking densities, altering feeding procedures, and making better decisions about where to put fish cages. It is important to recognize these challenges, not as excuses to do nothing, but as signposts about where to direct attention.

THE FISH DECADE

In spite of all the challenges and setbacks, there are good reasons to think the 2030s will be the decade of fish welfare policy. EFSA is scheduled to release scientific opinions on fish welfare standards during the current EU legislative term, which runs to 2029 – although it has been noted that 'the Commission seems to have fallen behind schedule, since EFSA's detailed mandate for the scientific opinion on salmon was slated for publication in June 2024.'[433] Depending on timing, it's possible the release of those opinions will align with the EU Council presidencies of major fish farming countries like Italy, Spain, Greece, or Cyprus. These would provide ideal moments for the Commission to propose welfare legislation, referencing EFSA opinions to ensure legislative proposals are keeping up with the latest animal welfare science.

First steps, like transitioning to humane slaughter, would usefully set a foundation for future reforms. From there, the EU should find that a structured and impactful approach to improving fish welfare is achievable, especially if it chooses to bring to life the actions outlined above, leverage public funding, and build on national legislation and economic feasibility data. For many areas of fish suffering, there are still a range of unknowns. But if the EU were to meet the challenge by identifying solutions and implementing them, it could take a leadership position on farmed fish welfare.

By embracing this role, the EU can ensure that fish welfare is not left to fragmented national initiatives, inconsistent enforcement, or vague voluntary guidelines. Instead, it can become part of a coherent European vision for humane food systems, one that includes all animals, whether they walk, fly, or swim.

CHAPTER 8

OVERCOMING THE BARRIERS
TO PROGRESS

One of the great advantages of EU-level legislation is that a single law can protect millions of animals in one fell swoop. In this sense, no other country, or group of countries, has done so much for so many animals. While many of the achievements of EU law have been built upon precedents set in member states, it is the EU system that means the impact for animals is greater than the sum of its parts.

The most impressive EU laws protecting farmed animals have been the ones that specifically target the worst practices of industrial animal agriculture: those that restrict the use of narrow crates for veal calves and individual stalls for sows;[434] ban conventional battery cages for egg-laying hens;[435] mandate methods of production labels for eggs;[436] and set limits on how many broiler chickens can be in one barn.[437] These were not just technical directives. These laws were also moral declarations that the EU would not sacrifice its values on the altar of industrial overreach.

It is to these laws that the EU owes its reputation as a global leader in the animal protection space. For decades, between the 1980s and the late 2000s, it advanced higher farmed animal standards by providing legislation that, if not perfect, at least responded to citizens' wishes, market trends, and the priorities of member states by concretely improving the lives of chickens, pigs, and cows. But leadership is not a permanent state. It must be continuously earned and renewed. As this book has shown, the EU's leadership position has ebbed, flowed, and even stilled, most notably after lawmakers failed even to put forward the 2023 farmed animal reform package. In the current 2024 to 2029 legislative term, the EU has a chance to reclaim its position as a leader by bringing its farmed animal protection standards into line with the latest science, market trends, citizen expectations, and member-state best practices. In doing so, it should ensure that it is creating policies that reward and support producers who keep animals more humanely.

SO WHY DON'T THEY DO IT?

Anyone who witnessed the failure of the 2023 reform package might fairly ask why lawmakers have not already done this. The short answer is some of them have. At national level, various policymakers have been pushing protections forward.

As described earlier in this book, member states like Austria, Czechia, Denmark, and Germany have continued to be at the forefront of ending the use of cages for egg-laying hens. In October 2024 Romania passed a law to phase out fur farming. In doing so they joined the majority of EU countries that have laws phasing out, or massively restricting, this production,[438] and answered the call of 1.5 million EU citizens who signed an ECI petition in 2022 to ban fur farming.[439]

In other positive developments, the Flanders region in Belgium has adopted maximum stocking density and enrichment standards for farmed turkeys,[440] creating a solid foundation for the EU's first turkey-specific protections in future. Germany, France, Italy, and Austria have banned or largely limited the killing of day-old male chicks by harnessing technological developments that allow farmers to identify the sex of the chick in the egg before it hatches. Germany has introduced production-method labelling for all fresh pork products, giving consumers their second credible higher-welfare product signal (eggs were the first). Malta adopted a law preventing the force-feeding of geese, expanding the number of EU members that have fully banned the practice to twenty-two.[441] And there are many other improvements making incremental steps towards a more compassionate union.[442] At EU level, a new organic regulation came into force in 2021, extending protections to farmed fish and rabbits, and to breeding pigs, chickens, and hens – the mothers of the pigs, chickens, and hens we farm for meat and eggs. The same regulation sets a limit on the number of animals per farm in broiler and egg production.[443]

As good as this news is, organic production remains a small share of the EU's farmed animals and the new regulations cannot replace the far more ambitious 2023 reform package the administration was courageously pursuing with the support of MEPs including Anja Hazekamp, Manuela Ripa, Michal Wiezik, Thomas Waitz, and Tilly Metz. These are MEPs who passed resolutions in favour of ending extreme confinement and conducted reports and inquiries into a range of abuses, especially the horrific conditions endured by animals during transport. These politicians will be seen, in the long arc of history, as those who led the charge for better EU leadership on animal welfare.

LETHARGY AND OPPOSING FORCES

Unfortunately, maintaining the status quo is usually easier than leading. Entire production systems, supply chain relationships, farmer training modules, government agencies, and financial instruments are established on the basis of one set of incentives and disincentives, and it is a rare occurrence that a proposed change offers a benefit across all dimensions at once. This is why, as I have argued in this book, the EU is best placed to act where the tide is already coming in, for example by phasing out cages for egg-laying hens.

The other drag on welfare legislation is that almost a quarter of the European Parliament's agriculture committee, and many former agriculture commissioners, are involved in the business of animal farming.[444] That means the tilt is toward minimum welfare with maximum production, rather than a production that maximizes or even protects welfare. This is not a slight against farmers. They clearly have the wisdom needed to realize higher welfare on farms, as per scientific recommendations, but the system is not set up to reward them for doing so.

On top of that, the EU does not see the welfare of farmed animals as an end goal in itself. Rather, it is seen as a component of consumer and food safety, or, in other words, a secondary concern. This is evident from the fact that legislative files on animal protection originate from the Directorate-General in charge of human health and food safety, DG SANTE, not the one in charge of agriculture – never mind that we don't have a specific DG for animal welfare, just a commissioner with 'animal welfare' in the title.

Nor is farmed animal cruelty top of mind for most citizens, partly because it is hidden from view and partly because citizens and politicians have not yet effectively organized animal welfare issues in a way that would create significant voting blocks – blocks that could credibly affect politicians' chances of getting into office. To the extent political leaders have taken a stand on this issue, they have tended to fall more on the Left/Green side of the EU political spectrum, despite many conservative governments, parties, and politicians actually supporting farmed animal protections.[445] This means animals lose out when political tides change. However, it also means there is room for cross-party support if MEPs are brave enough to work with their colleagues whose opinions differ on a range of other issues, but are united in opposing animal cruelty.

This lack of political organization means the widespread support for laws that improve welfare, as shown in Eurobarometer surveys and the 1.6 million people who signed the End the Cage Age petition,[446] remains somewhat hidden, suggesting a block of latent support is just waiting to be fired up. And when the public does rise up, politicians tend to take note. This latent support also raises a question for politicians: do EU leaders want to be purely reactive, i.e., waiting to act only where there is a voter backlash? Or do they want to proactively court those who speak up for animals by taking a stand as leaders in the field?

Powerful farm lobbies are another key factor in delaying, diluting, or stopping new protections for farmed animals. They achieve impressive feats and, despite including many wealthy farming corporations and farms, also receive large government subsidies,[447] meaning they have plenty of money to spend. The EU in effect pays to get lobbied by these interests.

Moreover, small farmers and family farms, so attractive to a public that wants to believe in rolling green fields and happy animals, are co-opted by the lobbies in different ways in order to sway public opinion. For example, COPA-COGECA, an active participant in the European Livestock Voice campaign effort,[448] underpins its demands by saying it represents over 22 million farmers and their family members. But that is not the number of farmers who are members of COPA-COGECA, small or otherwise; it is the number of Europeans who work in food production.[449] Such groups have been successful too at hijacking the legitimate anger of smaller farmers away from their actual problems, which include falling agricultural incomes and inflation,[450] toward destroying proposed EU policies – policies that might dent the incomes of agribusiness corporations by forcing them to adopt the higher-welfare practices many smaller farmers want to find a market for. Other tactics that can derail animal welfare legislation include funding shoddy science with the aim of questioning scientific opinions issued by EFSA,[451] opinions that have traditionally found that conventional animal farming practices fall far from offering best-in-class protections.

Part of the problem is that the scale of this moral atrocity is just too large for the human mind to bear: the suffering of the individual gets lost, and the only numbers that mean anything are the financial ones. Lewis Bollard, who leads Open Philanthropy's strategy for reducing the suffering of farmed animals, points to the words of Ruth Harrison, founder of the modern farmed animal welfare movement and author of the groundbreaking book *Animal Machines*:

> If one person is unkind to an animal it is considered to be cruelty, but where a lot of people are unkind to a lot of animals, especially in the name of commerce, the cruelty is condoned and, once large sums of money are at stake, will be defended to the last by otherwise intelligent people.[452]

Writing from across the Atlantic, the *New York Times* writer Nicholas Kristof makes a similar point when reflecting on videos made of the farming of pigs:

> If a teenage boy were to cut off the tails of animals and yank off their testicles, he might be arrested and castigated for his cruelty; if he grows up and becomes C.E.O. of a company that does this on a mass scale, he will get rich and be praised for his business acumen ... What I am confident of is that right now we're on the wrong side of history and that future generations will look back at videos like these and be baffled that nice people like us could blindly tolerate such systematized cruelty toward intelligent if cantankerous fellow mammals not so different from us.[453]

The good thing is that for problems of such vast and intractable scale where the public wants change but greater forces oppose it, we know what to do: look to the EU to lead, as it has done before. And if it does, what could this new leadership look like?

PILLARS OF RENEWED LEADERSHIP

While there are plenty of policies that would help animals, EU lawmakers would be justified in focusing on the easy wins, at least in the near term. Easy wins are ones that are already supported by the latest science and EU citizens, fit existing policy paradigms, and are already being worked on by the Commission and member states.

As this book has explained, the easiest wins for animals and policymakers would be phasing out extreme confinement for farmed animals, ending the risk of lower-welfare imports, and setting the foundations for future fish protections. The cage ban for laying hens is the EU's most immediate opportunity. A ban on cages fits because demand for cage-free eggs already exists, and several countries, including Austria, Czechia, Denmark, Germany, and Slovenia, have phased out or are planning to phase out cages. With market and political conditions in member states aligned, it is completely possible and legitimate to expect lawmakers to act. Plus, as mentioned, the Commission has already prepared much of the proposed legislation in the background, meaning there is every reason to include it in the Commission's 2026 work program.[454] This is not a question of whether, but when, and European leadership means ensuring that when is now, not after another decade's delay. We can expect negotiations, compromises, and amendments as legislators weigh what both the EU's egg industry and its citizens consider workable and acceptable.

This takes time, but it doesn't need to take another legislative term to make such straightforward progress. The exact transition times, the exact financial supports, the exact flock sizes this affects – all these have to be finalized, and not everyone will be happy with the proposals. But the important thing is that a new generation of hens would know nothing but a life outside of cages. As Kristof says in another article in the *New York Times*, there is no need for 'a misplaced nostalgia for traditional farming practices, just a pragmatic acknowledgment of animal suffering and trade-offs to reduce it.' And, he adds, while many of us 'aren't quite sure what rights animals should have, or how far to take this concern for animal well-being' we are 'learning as we go' and 'most are willing to pay a bit more to avoid torturing animals.'[455] Factory farming hurts farmers as much as it does animals. This is why the farmers are finding common ground with welfare NGOs in battling a system that has turned animals into living machines and farmers into serfs.[456]

Once that momentum has been recaptured, there's little reason to continue the extreme confinement of other species – like pigs, rabbits, and different birds – for which there has been similar scientific progress. Better yet, the EU should move to free broiler chickens from the shackles of their monstrous bodies by prohibiting the fastest-growing breeds and reducing how many can be crammed into one barn.

European research institutions continue to lead the world in animal cognition and welfare science. EFSA's scientific opinions provide the gold standard for welfare assessment globally. New scientific reports from EFSA on farmed fish welfare, which will build on previous findings in the 2000s, could help lay the terms of debate for applying existing protections appropriately to aquatic animals. Even without fresh EFSA reports, the existing evidence should provide enough material to support more specific protections. The other factor in fish's favour is that, legally and politically speaking, going from nothing to something means there is a range of low-hanging fruit that will improve both fish welfare and farmer incomes.

True leadership will additionally require the EU to protect its welfare progress from being undermined. That means ensuring its welfare standards are not circumvented by lower-welfare imports, an action that will reverberate far beyond the EU's borders. If the EU leads, others will follow. If Brussels sets ambitious standards, global markets will adapt. If the EU demonstrates that economic prosperity and animal welfare can coexist, it will inspire progress across continents.

This matters because, from turkeys to fish dying in agony, millions of animals remain outside existing EU protections. Leadership means expanding the circle of concern, not limiting it to the most economically convenient cases.

In themselves, all three of these avenues – ending extreme confinement, filling gaps in species-specific protection laws, and import protections – are valid contenders for action by the EU's 2024 to 2034 administrations. They have scientific and consumer support, and the EU is already considering them. In terms of the political paradigm, they are a good fit, mainly because they build on precedent in existing laws, and on the idea that the EU should have harmonious standards for the sake of its citizens and traders.

Lastly, policymakers should accept that farmed animals matter. They matter to farmers, to consumers, to businesses, and to anyone who believes the EU should be an ethical leader. And they matter for reasons of democratic accountability, public health, sustainability, food safety, and, in the longer run, economic innovation and competitiveness.

WHAT BETTER WELFARE CAN DO FOR THE EU

To state the obvious, giving 149 million laying hens kept in cages more space to express their natural instincts has clear benefits for animals, but it is also right by the millions of citizens who believe it is wrong to treat sentient beings this way. The same applies to any other species we uncage. In a world with better welfare laws, fish would not die slow agonizing deaths from suffocation, and supermarkets would not be flooded with low-welfare imports that undercut domestic farmers and force many consumers to choose cruelty.

Doing this will not only improve lives for animals. It will lift a burden from consumers currently forced to decipher small-print labels and make impossible, complex choices every day about what sorts of animal proteins they can afford, morally and financially, to buy and eat. Instead, they might be able to reclaim a very simple and very basic consumer right: that of knowing the products on the shelves are only there because they meet legal and ethical norms that ensure that the animals those foods came from lived a reasonable life and were killed as quickly and painlessly as possible.

The European Union was built on the revolutionary idea that nations could choose cooperation over conflict, that shared values could overcome narrow interests, that moral progress was possible through democratic institutions. These same principles still ring out, calling on Europe to extend its protection to the most vulnerable. The policy precedents exist. The best-practice examples are homegrown. The EU itself has created the necessary knowledge. The public support is clear.

Across twenty-seven democracies, from the Baltic Tigers to the Alpine Federations, citizens are calling for action. This is the European demos speaking, not in twenty-four languages, but in one moral voice. The only question remaining is whether European leaders will choose to act or falter, to inspire or disappoint, to fulfil the promise of this generation's citizens or pass the buck to the next.

The choice is clear. The time is now. This is Europe's moment to reclaim its moral leadership, not just in protecting the democratic rights and civil liberties of its citizens, but in protecting those non-human animals in its care.

ENDNOTES

CHAPTER 1: AN INTRODUCTION

1. EU Reporter Correspondent, '#EndTheCageAge: NGOs, MEPs and EU Citizens Unite to Celebrate Success of the European Citizens' Initiative (ECI),' Eureporter, October 8, 2019, https://www.eureporter.co/frontpage/2019/10/08/endthecageage-ngos-meps-and-eu-citizens-unite-to-celebrate-success-of-the-european-citizens-initiative-eci/.

2. Donald M. Broom, Animal Welfare in the European Union:Study for the PETI Committee (European Union, 2017), https://www.europarl.europa.eu/RegData/etudes/STUD/2017/583114/IPOL_STU(2017)583114_EN.pdf.

3. AGtivist Agency, 'The Face of European Farming,', https://stories.agtivistagency.com/the-face-of-european-farming/;
Sandra Laville and Helena Horton, 'Revealed: More than 24,000 Factory Farms Have Opened across Europe,' Guardian, June 12, 2025, https://www.theguardian.com/environment/2025/jun/12/research-reveals-24000-megafarms-across-europe;
AGtivist Agency, 'Megafarm Europe Interactive Map,', https://megafarms-europe.netlify.app/index.html;
Eurogroup for Animals, 'The Rise of Mega-Farms,' press release, June 12, 2025, https://www.eurogroupforanimals.org/news/rise-mega-farms-how-industrial-agriculture-taking-over-europe.

4. Guardian Staff, 'China's 26-Storey Pig Skyscraper Ready to Produce 1 Million Pigs a Year,' Guardian, November 25, 2022, https://www.theguardian.com/environment/2022/nov/25/chinas-26-storey-pig-skyscraper-ready-to-produce-1-million-pigs-a-year.

5. Council of the European Union, Council Directive 1999/74/EC of 19 July 1999 laying down minimum standards for the protection of laying Hens, consolidated version of December 14, 2019, Official Journal of the European Communities L 203, August 3, 1999, 53–57, https://eur-lex.europa.eu/eli/dir/1999/74/oj/eng;
Council of the European Union, Council Directive 2008/120/EC of 18 December 2008 laying down minimum standards for the protection of pigs, Official Journal of the European Union L 47, February 18, 2009, 5–13, https://eur-lex.europa.eu/legal-content/EN/TXT/PDF/?uri=CELEX:32008L0120;

Council of the European Union, Council Directive 97/2/EC of 20 January 1997 amending Directive 91/629/EEC laying down minimum standards for the protection of calves, Official Journal of the European Communities L 25, January 28, 1997, 24–25, https://eur-lex.europa.eu/eli/dir/1997/2/oj/eng;
Council of the European Union, Council Directive 2007/43/EC of 28 June 2007 laying down minimum rules for the protection of chickens kept for meat production, Official Journal of the European Union L 182, July 12, 2007, 19–28, https://eur-lex.europa.eu/eli/dir/2007/43/oj/eng.

6. For science compiled by the EU's European Food and Safety Authority (EFSA), see Søren Saxmose Nielsen et al., 'Welfare of Laying Hens on Farm,' EFSA Journal 21, no. 2 (February 1, 2023), https://doi.org/10.2903/j.efsa.2023.7789.

7. California Proposition 12, Farm Animal Confinement Initiative (2018), Ballotpedia, https://ballotpedia.org/California_Proposition_12%2C_Farm_Animal_Confinement_Initiative_%282018%29;
Lewis Bollard, 'A Big Supreme Court Win for Farm Animals,' Open Philanthropy Farm Animal Welfare Newsletter, May 12, 2023, https://farmanimalwelfare.substack.com/p/a-big-supreme-court-win-for-farm.

8. Philip Blenkinsop, 'Draghi Urges EU to Catch Up Rivals or Face "Slow Agony,"' Reuters, September 9, 2024, https://www.reuters.com/markets/europe/draghi-urges-reform-massive-investment-revive-lagging-eu-economy-2024-09-09/.

9. Alison Mood et al., 'Estimating Global Numbers of Farmed Fishes Killed for Food Annually from 1990 to 2019,' Animal Welfare 32 (2023), e12, https://doi.org/10.1017/awf.2023.4.

10. European Commission, Special Eurobarometer 533: Attitudes of Europeans towards Animal Welfare, (October 2023), https://europa.eu/eurobarometer/surveys/detail/2996.

11. National Museum of Asian Art, Smithsonian Institution, 'Neolithic Period (c. 7000–1700 B.C.E.), An Introduction,' Smarthistory, https://smarthistory.org/neolithic-period-china-introduction/.

12. 'Hunger in the Netherlands,' The Borgen Project, December 7, 2024, https://borgenproject.org/hunger-in-the-netherlands/.

13. Paul Tullis, 'Nitrogen Wars: The Dutch Farmers' Revolt That Turned a Nation Upside Down,' Guardian, November 16, 2023, https://www.theguardian.com/environment/2023/nov/16/nitrogen-wars-the-dutch-farmers-revolt-that-turned-a-nation-upside-down;
Laura Reiley, 'Cutting-Edge Tech Made This Tiny Country a Major Exporter of Food,' Washington Post, November 21, 2022, https://www.washingtonpost.com/business/interactive/2022/netherlands-agriculture-technology/.

14. Hannah Ritchie et al., 'Meat and Dairy Production,' Our World in Data, last updated December 2023, https://ourworldindata.org/meat-production.

15. Mood et al., 'Estimating Global Numbers of Farmed Fishes Killed for Food.'

16. European Commission, 'Eggs – Market Situation – Dashboard,' Directorate-General for Agriculture and Rural Development, last updated September 10, 2025, https://agriculture.ec.europa.eu/document/download/9bdf9842-1eb6-41a2-8845-49738b812b2b_en.

17. Global Ag Media, 'EU Milk Production Continues Downward Trend – GAIN,' The Dairy Site, June 6, 2024, https://www.thedairysite.com/news/eu-milk-production-continues-downward-trend-gain.

18. Sam Ducourant, 'Science or Ignorance of Animal Welfare? A Case Study: Scientific Reports Published in Preparation for the First European Directive on Animal Welfare (1979–1980),' Science, Technology, & Human Values 48, no. 1 (August 31, 2023): 139–66, https://doi.org/10.1177/01622439211040179.

19. European Union, Consolidated version of the Treaty on the Functioning of the European Union, art. 13, Official Journal of the European Union C 202, June 7, 2016, 54, https://eur-lex.europa.eu/eli/treaty/tfeu_2016/art_13/oj/eng.

20. 'Pigs Still Boiled Alive Despite Promises to Eradicate Practice,' DutchNews, June 29, 2020, https://www.dutchnews.nl/2020/06/pigs-still-boiled-alive-despite-promises-to-eradicate-practice-varkens-in-nood/;
Lucie Aubourg, 'This Horrific Video of Animal Cruelty Led to the Shutdown of a French Slaughterhouse,' Vice, October 15, 2015, https://www.vice.com/en/article/this-horrific-video-of-animal-cruelty-led-to-the-shutdown-of-a-french-slaughterhouse/;
Joaquim Elcacho, 'Matadero inhumano: nuevo vídeo denuncia de maltrato a corderos, vacas y caballos' [Inhumane Slaughterhouse: New Video Exposes Abuse of Lambs, Cows, and Horses], La Vanguardia, December 10, 2019, https://www.lavanguardia.com/natural/20191210/472161697537/dia-internacional-derechos-animales-video-denuncia-maltrato-matadero-albacete-equalia.html;
Animal Equality Italy, 'Sgozzati ancora coscienti: Animal Equality rivela le illegalità all'interno di un macello di maiali a Cremona' [Still Conscious, Slaughtered: Animal Equality Reveals Illegalities inside a Pig Slaughterhouse in Cremona], October 25, 2024, https://animalequality.it/comunicato-stampa/sgozzati-ancora-coscienti-animal-equality-rivela-le-illegalita-allinterno-di-un-macello-di-maiali-a-cremona/.

21. Bartosz Brzeziński and Max Griera, 'European Parliament Flinches at Factory Farming Reality,' Politico, March 25, 2024, https://www.politico.eu/article/european-parliament-flinches-at-factory-farming-reality/.

22. R.F. Wideman Jr. et al., A Growing Problem: Selective Breeding in the Chicken Industry;
The Case for Slower Growth (ASPCA, November 2015), https://www.aspca.org/sites/default/files/chix_white_paper_nov2015_lores.pdf.

23. Merete Forseth et al., 'Mortality Risk on Farm and During Transport: A Comparison of 2 Broiler Hybrids With Different Growth Rates,' Poultry Science 103, no. 3 (March 2024), 103395, https://doi.org/10.1016/j.psj.2023.103395.

24. Marc Bracke et al., Broiler Chickens: A Growing Problem (Eurogroup for Animals, November 2020), https://www.eurogroupforanimals.org/files/eurogroupforanimals/2021-12/2020_11_19_eurogroup_for_animals_broiler_report.pdf.

25. Bracke et al., Broiler Chickens.

26. Angela Symons, 'Egg-laying Hens are Killed after Just 18 Months. This Charity Gives Them a Brighter Future,' Euronews, January 2, 2024, https://www.euronews.com/green/2024/01/02/dogs-with-feathers-could-your-next-pet-be-a-chicken-rescue-from-an-egg-farm;
Four Paws, 'Life Expectancy of Chickens,' March 13, 2025, https://www.fourpaws.org/campaigns-topics/topics/farm-animals/life-expectancy-of-chickens.

27. National Pork Board, Group Housing Systems: Nutritional Considerations, https://porkcheckoff.org/wp-content/uploads/2021/05/Group-Housing-Systems-Nutritional-Considerations.pdf.

28. Sophie Kevany, 'More Than 1m Farmed Salmon Die at Supplier to Leading UK Retailers,' Guardian, October 22, 2024, https://www.theguardian.com/environment/2024/oct/22/more-than-1m-farmed-salmon-die-at-supplier-to-leading-uk-retailers.

29. Nicola Rotari, 'Treviso, pesci maltrattati nell'allevamento. "Sbattuti a terra e uccisi per asfissia, sofferenze inutili"' [Treviso, Mistreated Fish on the Farm: 'Slammed to the Ground and Killed by Asphyxiation, Useless Suffering'], Corriere del Veneto, April 30, 2024, https://corrieredelveneto.corriere.it/notizie/treviso/cronaca/24_aprile_30/treviso-pesci-maltrattati-nell-allevamento-sbattuti-a-terra-e-uccisi-per-asfissia-sofferenze-inutili-0770dd5a-03ff-4eca-bc18-e1409ae6dxlk.shtml;
Compassion in World Farming (CIWF), Investigation into EU Fish Farming and Slaughter: The Truth Behind the Fish Sold in Our Supermarkets (Compassion in World Farming, n.d.), https://www.ciwf.org.uk/media/7435552/70453_investigation-report-fish-appeal_-compassion-in-world-farming.pdf;
Animal Equality UK, 'New Investigation Shows Fish Killed While Fully Conscious in Scottish Salmon Slaughterhouse,' Eurogroup for Animals, February 15, 2021, https://www.eurogroupforanimals.org/news/new-investigation-shows-fish-killed-while-fully-conscious-scottish-salmon-slaughterhouse.

30. For detailed estimates see, Sagar Shah, Prospective Cost-Effectiveness of Farmed Fish Stunning Corporate Commitments in Europe, Rethink Priorities, March 14, 2024, https://rethinkpriorities.org/research-area/farmed-fish-corporate-commitments/.

31. Annick Hus and Steven P. McCulloch, 'The Political Salience of Animal Protection in the Netherlands (2012–2021) and Belgium (2010–2019): What Do Dutch and Belgian Political Parties Pledge on Animal Welfare and Wildlife Conservation?,' Journal of Agricultural and Environmental Ethics 36, art. 4 (2023), https://doi.org/10.1007/s10806-023-09899-6;
Paul Chaney et al., 'Sentience and Salience – Exploring the Party Politicization of Animal Welfare in Multi-Level Electoral Systems: Analysis of Manifesto Discourse in UK Meso Elections 1998–2017,' Regional & Federal Studies 32, no. 1 (2022): 115–40, https://doi.org/10.1080/13597566.2020.1853105; Colette S. Vogeler, 'Politicizing Farm Animal Welfare: A Comparative Study of Policy Change in the United States of America,' Journal of Comparative Policy Analysis: Research and Practice 23, no. 5–6 (2021): 526–43, https://doi.org/10.1080/13876988.2020.1742069;
Colette S. Vogeler, 'Why Do Farm Animal Welfare Regulations Vary between EU Member States? A Comparative Analysis of Societal and Party Political Determinants in France, Germany, Italy, Spain and the UK,' JCMS: Journal of Common Market Studies 57, no. 2 (2019): 317–35, https://doi.org/10.1111/jcms.12794;
Amelia Cornish et al., 'What We Know About the Public's Level of Concern for Farm Animal Welfare in Food Production in Developed Countries,' Animals 6, no. 11 (2016), 74, https://doi.org/10.3390/ani6110074;
Paul Chaney, 'Public Policy for Non-humans: Exploring UK State-wide Parties' Formative Policy Record on Animal Welfare, 1979–2010,' Parliamentary Affairs 67, no. 4 (2014): 907–34, https://doi.org/10.1093/pa/gss108.

32. Robin Hanson, 'To Oppose Polarization, Tug Sideways,' Overcoming Bias, March 13, 2019, https://www.overcomingbias.com/p/tug-sidewayshtml.

33. Lukas Paul Fesenfeld, The Political Economy of Food and Meat System Transformation (ETH Zurich, January 1, 2024), https://doi.org/10.3929/ethz-b-000648847.

CHAPTER 2: THE BIRTH OF FARMED
ANIMAL PROTECTION IN THE EU

34. Ruth Harrison, Animal Machines (Vincent Stuart Publishers, 1964), https://archive.org/details/animalmachines0000harr.

35. Kathleen Hall, 'Margaret Cooper Obituary,' Guardian, October 16, 2014, https://www.theguardian.com/theguardian/2014/oct/16/margaret-cooper.

36. Claas Kirchhelle, Bearing Witness: Ruth Harrison and British Farm Animal Welfare (1920-2000), (Palgrave Macmillan, 2021), https://link.springer.com/book/10.1007/978-3-030-62792-8.

37. Linda Lear, 'About Rachel Carson,' RachelCarson, https://www.rachelcarson.org/.

38. Technical Committee to Enquire into the Welfare of Animals Kept Under Intensive Livestock Husbandry Systems, Report of the Technical Committee to Enquire Into the Welfare of Animals Kept Under Intensive Livestock Husbandry Systems, (Her Majesty's Stationery Office, 1965), https://archive.org/details/b3217276x.

39. Farm Animal Welfare Council, 'Five Freedoms,' UK National Archives, archived October 12, 2012, https://webarchive.nationalarchives.gov.uk/ukgwa/20121010012427/http:/www.fawc.org.uk/freedoms.htm.

40. Swedish Institute, 'Astrid Lindgren: A Voice to Be Reckoned With,' last updated September 26, 2024, https://sweden.se/life/people/astrid-lindgren-a-voice-to-be-reckoned-with.

41. Karin Dirke, '11 Happy Cows? Unravelling Contexts of Swedish Farmed Animals,' in Animal Industries: Nordic Perspectives on the Exploitation of Animals since 1860, ed. Taina Syrjämaa et al. (De Gruyter, 2024), 213–228, https://doi.org/10.1515/9783110787337-015.

42. Svenska kyrkan , Församlingsutskottets betänkande 1995:501: Djuretik [Parish Committee Report 1995: 501: Animal Ethics], https://km.svenskakyrkan.se/km_om_95/of501.htm.

43. See references in Kirchhelle, Bearing Witness and D.G.M. Wood-Gush et al., 'Social Stress and Welfare Problems in Agricultural Animals,' in Behaviour of Domestic Animals, ed. ESE Hafez (Baillière Tindall, 1975), 182–200,

44. Birgitta Schwartz, 'The Animal Welfare Battle: The Production of Affected Ignorance in the Swedish Meat Industry Debate,' Culture and Organization 26, no. 1 (2020): 75–95, https://doi.org/10.1080/14759551.2018.1513937.

45. Małgorzata Szynkielewska, 'Astrid Lindgren: Storyteller of Childhoods,' Europeana, https://www.europeana.eu/en/stories/astrid-lindgren-storyteller-of-childhoods.

46. Association of Lawyers for Animal Welfare, The Farm Animal Welfare Council (FAWC) Annual Reviews, (December 2023), https://www.alaw.org.uk/wp-content/uploads/2023/12/The-Farm-Animal-Welfare-Council-FAWC.pdf.

47. Bundesministerium der Justiz und für Verbraucherschutz, 'Tierschutzgesetz § 16b' [Animal Welfare Act § 16b], https://www.gesetze-im-internet.de/tierschg/__16b.html; Deutscher Bundestag , Tierschutzbericht 2001: Bericht über den Stand der Entwicklung des Tierschutzes [Animal Welfare Report 2001: Report on the State of Development of Animal Welfare], Drucksache 14/5712, March 29, 2001, https://dserver.bundestag.de/btd/14/057/1405712.pdf.

48. Raad voor Dierenaangelegenheden, 'Geschiedenis van de RDA' [History of the RDA], https://www.rda.nl/over-ons/geschiedenis.

49. For those not already versed in the dizzying array of similarly named institutions in Europe, the Council of Europe is an international organisation focused on human rights, democracy, and the rule of law across the broader European continent, while the European Economic Community is the forerunner of what became the European Union (which includes bodies called the Council of the European Union and The European Council, also known as the Council of Member States' Ministers and Council of Member States' leaders).

50. Michael C. Appleby, 'The EU Ban on Battery Cages: History and Prospects,' in The State of the Animals II: 2003, eds. Deborah J. Salem and Andrew N. Rowan (Humane Society Press, 2003), Open Philanthropy, https://www.wellbeingintlstudiesrepository.org/sota_2003/13/.

51. 'Eurogroup for Animals,' Wikipedia, last modified November 7, 2024, https://en.wikipedia.org/wiki/Eurogroup_for_Animals.

52. 'What We Do,' Intergroup on the Welfare & Conservation of Animals, https://www.animalwelfareintergroup.eu/what-we-do.

53. Carolina Maciel and Bettina Bock, 'Modern Politics in Animal Welfare? The Changing Character of European Animal Welfare Governance and the Role of Private Standards,' International Journal of Sociology of Agriculture and Food 20, no. 2 (2013): 219–35, https://www.researchgate.net/publication/257934830.

54. P. Alarcon et al., 'Classical BSE in Great Britain: Review of Its Epidemic, Risk Factors, Policy and Impact,' Food Control 146 (2023): 109490, https://doi.org/10.1016/j.foodcont.2022.109490.

55. National CJD Research & Surveillance Unit, Table Showing Creutzfeldt-Jakob Disease in the UK by Calendar Year, last updated March 7, 2022,

56. Department of Agriculture, Environment and Rural Affairs (DAERA), 'BSE Feed Controls,', https://www.daera-ni.gov.uk/articles/bse-feed-controls.

57. Alarcon et al., 'Classical BSE in Great Britain.'

58. Compassion in World Farming (CIWF), 'Animal Sentience: The Highs and Lows,', https://www.ciwf.org.uk/our-campaigns/other-campaigns/animal-sentience-the-highs-and-lows/.

59. Beata Rojek, Animal Welfare Protection in the EU (European Parliamentary Research Service, May 2023), https://www.europarl.europa.eu/RegData/etudes/BRIE/2023/747131/EPRS_BRI(2023)747131_EN.pdf.

60. European Union, Treaty of Lisbon amending the Treaty on European Union and the Treaty Establishing the European Community, arts. 5b, 21, Official

Journal of the European Union C 306, December 17, 2007, https://eur-lex.europa.eu/legal-content/EN/TXT/PDF/?uri=OJ:C:2007:306:FULL.

61. European Union, Council Regulation (EC) No 1099/2009 of 24 September 2009 on the protection of animals at the time of killing (Text with EEA Relevance), Official Journal of the European Union L 303, November 18, 2009, 1–30, https://eur-lex.europa.eu/eli/reg/2009/1099/oj.

62. European Union, Council Directive 93/119/EC of 22 December 1993 on the protection of animals at the time of slaughter or killing, Official Journal of the European Communities L 340, December 31, 1993, 21–34, https://eur-lex.europa.eu/eli/dir/1993/119/oj.

63. European Union, Regulation (EU) 2016/429 of the European Parliament and of the Council of 9 March 2016 on transmissible animal diseases and amending and repealing certain acts in the area of animal health ('Animal Health Law'), Official Journal of the European Union L 84, March 31, 2016, 1–208, https://eur-lex.europa.eu/eli/reg/2016/429/oj;
European Union, Regulation (EU) 2017/625 of the European Parliament and of the Council of 15 March 2017 on official controls and other official activities performed to ensure the application of food and feed law, rules on animal health and welfare, plant health and plant protection products, amending Regulations (EC) No 999/2001, (EC) No 396/2005, (EC) No 1069/2009, (EC) No 1107/2009, (EU) No 1151/2012, (EU) No 652/2014, (EU) 2016/429 and (EU) 2016/2031 of the European Parliament and of the Council, Council Regulations (EC) No 1/2005 and (EC) No 1099/2009 and Council Directives 98/58/EC, 1999/74/EC, 2007/43/EC, 2008/119/EC and 2008/120/EC, and repealing Regulations (EC) No 854/2004 and (EC) No 882/2004 of the European Parliament and of the Council, Council Directives 89/608/EEC, 89/662/EEC, 90/425/EEC, 91/496/EEC, 96/23/EC, 96/93/EC and 97/78/EC and Council Decision 92/438/EEC (Official Controls Regulation), Official Journal of the European Union L 95, April 7, 2017, 1–142, https://eur-lex.europa.eu/eli/reg/2017/625/oj.

64. European Union, Council Directive 2007/43/EC of 28 June 2007 laying down minimum rules for the protection of chickens kept for meat production (Text with EEA Relevance), Official Journal of the European Union L 182, July 12, 2007, 19–28, https://eur-lex.europa.eu/eli/dir/2007/43/oj.

65. European Commission, EU Animal Welfare Strategy 2012–15 – Evaluation, https://ec.europa.eu/info/law/better-regulation/have-your-say/initiatives/2140-EU-animal-welfare-strategy-2012-15-evaluation_en.

66. Council of the European Union, Background Note: Evaluation of the EU Animal Welfare Strategy 2012–2015, https://www.consilium.europa.eu/media/49315/background-note-evaluation-of-the-eu-animal-welfare-strategy-2012-2015.pdf.

67. European Commission, Evaluation ('Fitness Check') of the EU Legislation on the Welfare of Farmed Animals (October 2022), https://food.ec.europa.eu/animals/animal-welfare/evaluations-and-impact-assessment/evaluation-fitness-check-eu-legislation-welfare-farmed-animals_en.

68. Geoff Tansey and Joyce D'Silva, eds., The Meat Business: Devouring a Hungry Planet (Earthscan Publications, 1999), https://www.routledge.com/The-Meat-Business-Devouring-a-Hungry-Planet/Tansey-DSilva/p/

book/9780367275990?srsltid=AfmBOor_fD3XODth8_auBn2RiaczoIvqofvc-8jiEzCov_63fM5VFxclp.

69. Colette S. Vogeler, 'Market-Based Governance in Farm Animal Welfare: A Comparative Analysis of Public and Private Policies in Germany and France,' *Animals* 9, no. 5 (2019): 267, https://doi.org/10.3390/ani9050267.

70. European Commission, Options for Animal Welfare Labelling and the Establishment of a European Network of Reference Centres for the Protection and Welfare of Animals, EUR-Lex 52009DC0584 (November 28, 2009),

71. Jane Dalton and Claire Colley, 'Cows on M&S and Müller Farms Kicked, Hit with Chains and Sworn At,' *Independent*, September 14, 2024, https://www.independent.co.uk/news/uk/home-news/cows-abuse-muller-marks-spencers-milk-rspca-b2611198.html.

72. Open Philanthropy, 'Cage-Free Reforms,', https://www.openphilanthropy.org/focus/farm-animal-welfare/cage-free-reforms/;
CIWF, 'What Is Better Chicken?,' Better Chicken, https://www.betterchicken.org.uk/.

73. 'The Better Life Label,' Beter Leven Dierenbescherming, https://beterleven.dierenbescherming.nl/english/.

74. Saatkamp Helmut et al., 'Transition from Conventional Broiler Meat to Meat from Production Concepts with Higher Animal Welfare: Experiences from The Netherlands,' *Animals* 9, no. 8 (2019), https://doi.org/10.3390/ani9080483.

75. Flora Southey, 'Chicken Sold in Dutch Supermarkets to Lead a "Better Life" by 2023: "This Is a Big Step Towards Better Animal Welfare,"' FoodNavigator, August 31, 2021, https://www.foodnavigator.com/Article/2021/08/31/Chicken-sold-in-Dutch-supermarkets-to-lead-a-better-life-by-2023-This-is-a-big-step-towards-better-animal-welfare/.

CHAPTER 3: WHEN THE EU LED THE WORLD: A BRIEF HISTORY OF SPECIES-SPECIFIC LAWS

76. Neil Dullaghan, Strategic Considerations for Upcoming EU Farmed Animal Legislation(Rethink Priorities, April 13, 2021), https://rethinkpriorities.org/research-area/strategic-considerations-for-upcoming-eu-farmed-animal-legislation;
Neil Dullaghan, Do Countries Comply with EU Animal Welfare Laws? (Rethink Priorities, August 16, 2020), https://rethinkpriorities.org/research-area/do-countries-comply-with-eu-animal-welfare-laws.

77. The act stated that animals must be properly treated and must not, by neglect, overstrain or in any other way, be subject to unnecessary suffering; furthermore, anyone keeping animals should see that they have sufficient and suitable food and drink, and that they are properly cared for in suitable accommodation. See Appleby, 'The EU Ban on Battery Cages.'

78. Franz Jessen, 'Egg on the Face of Government,' *Economic Affairs* 3, no. 1 (October 1982): 61–63, https://doi.org/10.1111/j.1468-0270.1982.tb01469.x.

79. Appleby, 'The EU Ban on Battery Cages';
Victoria Sandilands and Paul M. Hocking, eds., *Alternative Systems for Poultry: Health, Welfare and Productivity*, Poultry Science Symposium Series, vol. 30

(CABI Publishing, 2012), https://www.cabidigitallibrary.org/doi/book/10.1079/9781845938246.0000.

80. European Commission, Directorate-General for Health and Food Safety, 'Laying Hens,',

81. F. W. R. Brambell et al., Report of the Technical Committee to Enquire into the Welfare of Animals Kept under Intensive Livestock Husbandry Systems, Cmnd. 2836 (Her Majesty's Stationery Office, 1965), https://edepot.wur.nl/134379; Melissa Elischer, 'The Five Freedoms: A History Lesson in Animal Care and Welfare,' Michigan State University, September 6, 2019, https://www.canr.msu.edu/news/an_animal_welfare_history_lesson_on_the_five_freedoms; Lewis Bollard, 'What Would Ruth and Henry Do?,' Open Philanthropy, July 30, 2024, https://www.openphilanthropy.org/research/what-would-ruth-and-henry-do/.

82. Heinzpeter Studer, How Switzerland Got Rid of Battery Cages, trans. Anja Schmidtke (Pro Tier International, 2001), https://www.upc-online.org/battery_hens/SwissHens.pdf.

83. Council of Europe, European Convention for the Protection of Animals Kept for Farming Purposes, Treaty No. 87 (Strasbourg, March 10 1976), https://www.coe.int/en/web/conventions/full-list?module=treaty-detail&treatynum=087.

84. Ducourant, 'Science or Ignorance of Animal Welfare?'

85. Commission of the European Communities, Report from the Commission to the Council concerning the Keeping of Laying Hens in Cages: Proposal for a Council Directive, COM(81) S20 final, Brussels, August 3, 1981, https://aei.pitt.edu/50500/1/A10209.pdf.

86. Appleby, 'The EU Ban on Battery Cages.'

87. HC Debate, November 17, 1981, vol. 13, 'European Community Proposals (Battery Hens),' https://hansard.parliament.uk/commons/1981-11-17/debates/117eca6b-beaf-42f7-aad4-0f7dfe614acf/EuropeanCommunityProposals(BatteryHens).

88. Council of the European Communities, Opinion No. C343/81 on the proposal for a Council Directive laying down minimum standards for the protection of laying hens kept in battery cages, Official Journal of the European Communities C 343, December 31, 1981, 48–50, https://eur-lex.europa.eu/legal-content/EN/TXT/PDF/?uri=CELEX:51981AC1076&from=EN.

89. Council of the European Communities, Council Directive 86/113/EEC of 25 March 1986 laying down minimum standards for the protection of laying hens kept in battery cages, Official Journal of the European Communities L 95, April 10, 1986, 45–48, https://eur-lex.europa.eu/legal-content/EN/TXT/HTML/?uri=CELEX:31986L0113&from=EN.

90. Economic and Social Committee, Opinion on the 'Proposal for a Council Directive laying down minimum standards for the protection of laying hens kept in various systems of rearing,' Official Journal of the European Communities C 407, December 28, 1998, 214, https://eur-lex.europa.eu/legal-content/EN/TXT/HTML/?uri=CELEX%3A51998AC1155.

91. Studer, How Switzerland Got Rid of Battery Cages.

92. Studer, How Switzerland Got Rid of Battery Cages.

93. Agneta Brasch and Christer Nilsson, Sveriges omställning till alternativa inhysningssystem för värphöns – en tillbakablick [Sweden's Transition to Alterna-

119

tive Housing Systems for Laying Hens: A Review], Rapport 2008: 33(Jordbruks Verket, December 15, 2008), https://www2.jordbruksverket.se/webdav/files/SJV/trycksaker/Pdf_rapporter/ra08_33.pdf.

94. Steve Lohr, 'Swedish Farm Animals Get a Bill of Rights,' New York Times, October 25, 1988, https://www.nytimes.com/1988/10/25/world/swedish-farm-animals-get-a-bill-of-rights.html.

95. Animal Welfare Act (Sweden) (1988:534) 1988 (2007) , https://www.ecolex.org/details/legislation/animal-welfare-act-1988534-lex-faoc019544/.

96. United Poultry Concerns, 'Swedish Battery Hen Ban in Danger!' Poultry Press, Winter 1996/97, https://www.upc-online.org/w96swedish_battery.html; Lena Lindström, Hönan eller ägget: En genomlysning av äggindustrin [The Chicken or the Egg: A Review of the Egg Industry], (Djurens Rätt, 2009), https://djurensratt.se/sites/default/files/honan_eller_agget_0.pdf.

97. Brasch and Nilsson, Sveriges omställning.

98. Miljö- och jordbruksutskottet, Betänkande 1999/2000:MJU05, Livsmedelskontroll [Food Control Report 1999/2000:MJU05] (Sveriges Riksdag),

99. Tansey and D'Silva, The Meat Business.

100. Joel Mead, 'How Free-Range Eggs Became the Norm in Supermarkets – and Sold Customers a Lie,' University of Liverpool, March 13, 2023, https://news.liverpool.ac.uk/2023/03/13/how-free-range-eggs-became-the-norm-in-supermarkets-and-sold-customers-a-lie/.

101. Tansey and D'Silva, The Meat Business.

102. Mead, 'How Free-Range Eggs Became the Norm in Supermarkets'; British Lion Eggs, 'UK Egg Industry Data,' , https://www.egginfo.co.uk/egg-facts-and-figures/industry-information/data.

103. Bundesverfassungsgericht, Beschluss des Zweiten Senats vom 6. Juli 1999 – 2 BvV 3/90 [Judgment of July 6, 1999, 2 BvF 3/90], press release, https://www.bundesverfassungsgericht.de/SharedDocs/Entscheidungen/DE/1999/07/fs19990706_2bvf000390.html.

104. Lindström, Hönan eller ägget.

105. Food and Agriculture Organization of the United Nations), Ordinance by the Viennese government amending the Regional Ordinance on the keeping of battery chickens, 5(3) and (4)) https://www.fao.org/faolex/results/details/en/c/LEX-FAOC033474.

106. Verein gegen Tierfabriken (VGT), 'Animal Law:What VGT Has Achieved in Austria!',December 11, 2007, https://www.abolitionistapproach.com/media/links/p140/another-portion.pdf.

107. Commission of the European Communities, Communication from the Commission on the protection of laying hens kept in various systems of rearing, COM(1998) 135 final, March 11, 1998, https://aei.pitt.edu/10645/1/10645.pdf; Animal Welfare Decree (Finland) (No. 396/1996), Ministry of Agriculture and Forestry, June 27, 1996, last checked July 2015, https://www.animallaw.info/statute/finland-animal-welfare-decree.

108. Agra CEAS Consulting Ltd., Study on the Socio-Economic Implications of the Various Systems to Keep Laying Hens, final report for the European Commission,

December 2004, https://www.yumpu.com/en/document/view/42854993/2120-fi-nal-reportpdf-agra-ceas-consulting.

109. Taina Vesanto et al., Työryhmämuistio MMM 2003:14, Kananmunien tuotan-tostrategia: Uusiin tuotantomuotoihin siirtyminen vuonna 2012, Loppuraport-ti [Working Group Memorandum MMM 2003:14, Chicken Egg Production Strategy: Transition to New Forms of Production in 2012, Final Report], (Maa-ja metsätalousministeriö, 2003),

110. Productivity Commission, Battery Eggs Sale and Production in the ACT, (AusInfo, 1998), https://econwpa.ub.uni-muenchen.de/econ-wp/othr/pa-pers/0107/0107009.pdf.

111. United Kingdom, Foreign & Commonwealth Office, UK Presidency of the Coun-cil of Ministers of the European Union: Work Programme, archived July 31, 2003, https://web.archive.org/web/20030731055137/http://presid.fco.gov.uk/workprog/within.shtml.

112. Commission of the European Communities, COM(1998) 135 final, proposal for a Council Directive laying down minimum standards for the protection of lay-ing hens, Brussels, March 11, 1998, https://aei.pitt.edu/10645/1/10645.pdf.

113. Commission of the European Communities, COM(1998) 135 final, proposal for a Council Directive.

114. Robert Thomson et al., 'A New Dataset on Decision-Making in the European Union Before and After the 2004 and 2007 Enlargements (DEUII),' Journal of European Public Policy 19, no. 4 (2012): 604–22, https://doi.org/10.1080/13501 763.2012.662028.

115. Appleby, 'The EU Ban on Battery Cages';
TiHo Hannover, 'Wie die Legehenne den Käfig kam II' [How the Laying Hen got into the Cage II], archived July 31, 2021, https://web.archive.org/web/20210731024635/https://wing.tiho-hannover.de/forschung/kritische-the-men/wie-die-legehenne-den-kaefig-kam-ii.html.

116. 'Adolfo Sansolini,' Animal Welfare and Trade, November 15, 1966,

117. Bundesverfassungsgericht , Judgment of July 6, 1999, 2 BvF 3/90.

118. Council of the European Union, Council Directive 1999/74/EC of 19 July 1999 laying down minimum standards for the protection of laying hens, consolidated version of December 14, 2019, Official Journal of the European Communities L 203, August 3, 1999, 53–57, https://eur-lex.europa.eu/eli/dir/1999/74/oj/eng.

119. Studer, How Switzerland Got Rid of Battery Cages.

120. L. Bäckström, 'Environment and Animal Health in Piglet Production: A Field Study of Incidences and Correlations,' Acta Veterinaria Scandinavica Supple-mentum 41 (1973): 1–240, https://pubmed.ncbi.nlm.nih.gov/4518921/;
Marlene Halverson, Management in Swedish Deep-Bedded Swine Housing Sys-tems: Background and Behavioral Considerations (Institute for Agriculture and Trade Policy, 2001), https://p2infohouse.org/ref/21/20979.htm.

121. The Animal Welfare Ordinance (Sweden) (1988:539), Ministry of Agricul-ture, Food and Fisheries, December 19, 2003, https://faolex.fao.org/docs/pdf/swe19545E.pdf;
G. Ståhle, 'Ban on Stalls and Tethers for Sows:Swedish Experience,' paper given to Eurogroup for Animal Welfare at the European Parliament, Brussels, Federa-tion of Swedish Farmers [LRF], June 2000. Cited in Jacky Turner, The Welfare of

Europe's Sows in Close Confinement Stalls: A Report Prepared for the European Coalition for Farm Animals (CIWF, December 2011), https://www.ciwf.org.uk/media/3818886/welfare-of-europes-sows-in-close-confinement-stalls.pdf.

122. Gerard O'Dwyer, 'Swedish Pig Farmers Call for State Support,' Food Navigator, March 3, 2014, last updated March 18, 2017, https://www.foodnavigator.com/Article/2014/03/03/Swedish-government-seek-initiatives-to-tackle-high-cost-pig-farming/;
Organisation for Economic Co-operation and Development (OECD), Innovation, Agricultural Productivity and Sustainability in Sweden, OECD Food and Agricultural Reviews (OECD Publishing, 2018), https://www.oecd.org/content/dam/oecd/en/publications/reports/2018/06/innovation-agricultural-productivity-and-sustainability-in-sweden_g1g8b712/9789264085268-en.pdf.

123. HC Debate, January 25, 1991, vol. 184, 'Pig Husbandry Bill,' https://hansard.parliament.uk/commons/1991-01-25/debates/82a2d463-e92c-40e4-aff5-a5bce0591335/PigHusbandryBill;
CIWF, 'Obituary: Sir Richard Body 1927–2018,' October 1, 2018, https://www.ciwf.org.uk/news/obituary-sir-richard-body-1927-2018/;
Royal British Legion, 'The Gulf War,', https://www.britishlegion.org.uk/get-involved/remembrance/stories/the-gulf-war.

124. Animal Aid, Best of British? The Pig Industry Exposed (Animal Aid, 2008), https://web.archive.org/web/20250602031622/https://www.animalaid.org.uk/wp-content/uploads/2016/08/Pigreport.pdf;
The Welfare of Pigs Regulations 1991 (UK), S.I. 1991/1477 , legislation.gov.uk, https://www.legislation.gov.uk/uksi/1991/1477/made.

125. House of Commons Environment, Food and Rural Affairs Committee, The English Pig Industry: First Report of Session 2008–09, HC 96 (The Stationery Office, 2009), https://publications.parliament.uk/pa/cm200809/cmselect/cmenvfru/96/96.pdf.

126. Vincent ter Beek, 'UPDATE: British Demand Enforcement of EU Sow Stall Ban,' Pig Progress, March 21, 2012, https://www.pigprogress.net/pigs/update-british-demand-enforcement-of-eu-sow-stall-ban/.

127. European Commission, Proposal for a Council Regulation (EEC) concerning minimum standards for the protection of pigs kept in intensive farming systems, COM(89) 115 final (Presented by the Commission on June 23, 1989), Official Journal of the European Communities C 214, August 21, 1989, 5, https://eur-lex.europa.eu/legal-content/EN/TXT/PDF/?uri=CELEX:51989PC0115&from=EN.

128. Economic and Social Committee, 'Information and Notices,' Official Journal of the European Communities C 62, March 12, 1990, https://eur-lex.europa.eu/legal-content/EN/TXT/PDF/?uri=OJ:C:1990:062:FULL&from=EN;
European Economic and Social Committee, Opinion on the Proposal for a Council Directive laying down minimum rules for the protection of chickens kept for meat production (COM[2005] 221 final – 2005/0099 CNS), Official Journal of the European Union C 28, February 3, 2006, 25–28, https://eur-lex.europa.eu/legal-content/EN/TXT/?uri=uriserv:OJ.C_.2006.028.01.0025.01.ENG&toc=OJ:C:2006:028:TOC.

129. European Parliament, Minutes of Proceedings of the Sitting of Thursday, April 5, 1990, Official Journal of the European Communities C 113, May 7,

1990, 124, https://eur-lex.europa.eu/legal-content/EN/TXT/PDF/?uri=OJ:-JOC_1990_113_R_0124_01&from=EN.

130. Turner, The Welfare of Europe's Sows .

131. European Commission, Scientific Veterinary Committee, The Welfare of Intensively Kept Pigs, Doc XXIV/B3/ScVC/0005/1997 (European Commission, 1997), https://edepot.wur.nl/531629.

132. Mark Honeyman and Dennis Kent, 'Performance of Pigs in a Swedish Bedded Group Lactation and Nursery System,' Pig Site, February 26, 2001, https://www.thepigsite.com/articles/performance-of-pigs-in-a-swedish-bedded-group-lactation-and-nursery-system;
Peter H. Brooks, Group Housing of Sows: The European Experience (CABI, 2013), https://www.cabidigitallibrary.org/doi/pdf/10.5555/20133192165.

133. Herman M. Vermeer et al., 'Comparison of Group Housing Systems for Sows and Introduction in Practice in the Netherlands Driven by Legislation and Market,' paper presented at the Annual International Meeting Sponsored by ASAI, Toronto, July 18–21, 1999, https://www.researchgate.net/publication/259705446_COMPARISON_OF_GROUP_HOUSING_SYSTEMS_FOR_SOWS_AND_INTRODUCTION_IN_PRACTICE_IN_THE_NETHERLAND_DRIVEN_BY_LEGISLATION_AND_MARKET.

134. Eric van Heugten, 'Housing of Sows and Gilts in Denmark,' Pig Site, April 16, 2003, https://www.thepigsite.com/articles/housing-of-sows-and-gilts-in-denmark;
DanBred, 'Embrace Loose Housing of Sows: Danish Experiences with DanBred Genetics,' May 8, 2024, https://danbred.com/embrace-loose-housing-of-sows/.

135. Joel Novek, 'Pigs and People: Sociological Perspectives on the Discipline of Non-human Animals in Intensive Confinement,' Society & Animals 13, no. 3 (2005): 220–44, https://www.animalsandsociety.org/wp-content/uploads/2016/01/novek.pdf;
Turner, The Welfare of Europe's Sows.

136. M.B.M. Bracke et al., 'Overall Welfare Assessment of Pregnant Sow Housing Systems Based on Interviews with Experts,' Netherlands Journal of Agricultural Science 47, no. 2 (1999): 93–104, https://library.wur.nl/ojs/index.php/njas/article/view/469.

137. European Commission, Scientific Veterinary Committee, The Welfare of Intensively Kept Pigs.

138. Council of the European Union, Council Directive 2008/120/EC of 18 December 2008 laying down minimum standards for the protection of pigs, Official Journal of the European Union L 47, February 18, 2009, 5–13, https://eur-lex.europa.eu/legal-content/EN/TXT/PDF/?uri=CELEX:32008L0120.

139. David B. Kopel, 'Calf Cruelty Is Not Necessary,' Houston Post, April 24, 1989, reprinted in Congressional Record, July 25, 1989, E2670;
Animal Welfare Act (Sweden).

140. Ministry of Agriculture and Forestry, Finland, Animal Welfare Decree (396/1996, as amended up to 401/2006), unofficial translation, archived October 11, 2009, https://web.archive.org/web/20091011211819/http://www.finlex.fi/en/laki/kaannokset/1996/en19960396.pdf.

141. Joyce D'Silva, 'Peter Roberts,' Guardian, November 23, 2006, https://www.theguardian.com/environment/2006/nov/23/obituaries.animalrights; Tansey

and D'Silva, The Meat Business; Peter Singer, ed., In Defense of Animals: The Second Wave (Blackwell, 1985), https://grupojovenfl.wordpress.com/wp-content/uploads/2019/10/peter-singer-in-defense-of-animals-1.pdf; Carol McKenna, The Case Against the Veal Crate: An Examination of the Scientific Evidence That Led to the Banning of the Veal Crate System in the EU and of the Alternative Group Housed Systems That Are Better for Calves, Farmers and Consumers (CIWF, April 2001), https://www.ciwf.org.uk/media/3818635/case-against-the-veal-crate.pdf.

142. European Parliament, Minutes of Proceedings of the Sitting of Friday, 20 February 1987, Official Journal of the European Communities C 76, March 23, 1987, 158, https://eur-lex.europa.eu/legal-content/EN/TXT/PDF/?uri=OJ:JOC_1987_076_R_0158_01&from=EN.

143. HL Debate, November 19, 1987, vol. 490, cols. 350–64, 'Welfare of Battery Hens Regulations 1987,' https://api.parliament.uk/historic-hansard/lords/1987/nov/19/welfare-of-battery-hens-regulations-198.

144. European Commission, Proposal for a Council Regulation (EEC) concerning minimum standards for the protection of calves kept in intensive farming systems, Official Journal of the European Communities C 214, August 21, 1989, https://eur-lex.europa.eu/legal-content/EN/TXT/PDF/?uri=CELEX:51989P-C0114&from=EN.

145. Economic and Social Committee, 'Information and Notices.'

146. McKenna, The Case Against the Veal Crate;
Council of the European Communities, Council Directive 91/629/EEC of 19 November 1991 laying down minimum standards for the protection of calves, Official Journal of the European Communities L 340, December 11, 1991, 28–32, https://eur-lex.europa.eu/eli/dir/1991/629/oj/eng.

147. McKenna, The Case Against the Veal Crate;
Productschap Vee en Vlees, Jaarverslag 2001 [Inventory of the Archives of the Product Board for Livestock and Meat (PVV)] (Productschap Vee en Vlees, 2002), 24, https://www.nationaalarchief.nl/onderzoeken/archief/2.25.80.

148. McKenna, The Case Against the Veal Crate.

149. McKenna, The Case Against the Veal Crate.

150. HC Debate, June 12, 1991, vol. 192, col. 384, 'Welfare of Calves (Export),' https://hansard.parliament.uk/Commons/1991-06-12/debates/e2706d68-03b1-4870-bbe7-5bf99f323422/WelfareOfCalves(Export).

151. Myles Neligan, 'Public Morality at Stake in ECJ Ruling on Veal,' Politico, April 12, 2014, https://www.politico.eu/article/public-morality-at-stake-in-ecj-ruling-on-veal/.

152. Court of Justice of the European Union, The Queen v. Minister of Agriculture, Fisheries and Food, ex parte Compassion in World Farming Ltd., Case C-1/96, March 19, 1998, https://eur-lex.europa.eu/legal-content/EN/TXT/?uri=CELEX:61996CJ0001.

153. HC Debate, January 26, 1995, vol. 253, 'Oral Answers to Questions: Agriculture, Fisheries and Food—Rural White Paper,' https://hansard.parliament.uk/Commons/1995-01-26/debates/082eda2c-9db6-4aa2-b7f4-df97cbb4bc62/CommonsChamber;

HC Debate, February 22, 1995, vol. 255, 'Live Animals (Export),' https://api.parliament.uk/historic-hansard/commons/1995/feb/22/live-animals-export; HC Debate, February 23, 1995, 'Oral Answers to Questions: Agriculture, Fisheries and Food,' https://publications.parliament.uk/pa/cm199495/cmhansrd/1995-02-23/Orals-1.html.

154. Marie Woolf, 'EU Report Exposes Veal Myth: The Campaign Against Cruel Treatment of Calves Will Gain Strength from a Brussels Commission—but New Controversy Looms over Lamb,' Independent, January 15, 1995, https://www.independent.co.uk/news/eu-report-exposes-veal-myth-the-campaign-against-cruel-treatment-of-calves-will-gain-strength-from-a-brusse-ls-commission-but-new-controversy-looms-over-lamb-1568060.html.

155. HC Debate, February 22, 1995, 'Live Animals (Export).'

156. HC Debate, December 14, 1995, vol. 268, 'Veal Crates,' https://hansard.parliament.uk/Commons/1995-12-14/debates/2568e412-99de-4a02-a36c-6787a11ace92/VealCrates.

157. Commission of the European Communities, Communication from the Commission to the Council and the European Parliament on the welfare of calves, COM(95)711 final, December 15, 1995, https://eur-lex.europa.eu/LexUriServ/LexUriServ.do?uri=COM%3A1995%3A0711%3AFIN%3AEN%3APDF.

158. McKenna, The Case Against the Veal Crate.

159. European Commission, Proposal for a Council Directive amending Directive 91/629/EEC laying down minimum standards for the protection of calves, applying from 1 January 1998, COM(96) 21 final, January 24, 1996, https://eur-lex.europa.eu/legal-content/EN/TXT/PDF/?uri=CELEX:51996PC0021; McKenna, The Case Against the Veal Crate.

160. Council of the European Union, Council Directive 97/2/EC of 20 January 1997 amending Directive 91/629/EEC laying down minimum standards for the protection of calves, Official Journal of the European Communities L 25, January 28, 1997, 24–25, https://eur-lex.europa.eu/eli/dir/1997/2/oj/eng.

161. CIWF, 'Fast-growing birds trapped in oversized bodies can't behave like chickens,' Facebook, November 20, 2024, https://www.facebook.com/story.php?story_fbid=964264159062440&id=10499876484&_rdr.

162. C. Weeks and A. Butterworth, eds., Measuring and Auditing Broiler Welfare (CABI Publishing, 2004), https://www.scienceopen.com/book?vid=cd7a2416-58cf-4fec-ae7a-cfaabfb348ab.

163. Bekendtgørelse om beskyttelse af æglæggende høner [Order on the Protection of Laying Hens], LOV nr 1069 af 12/12/2001, Retsinformation, https://www.retsinformation.dk/eli/lta/2001/1069.

164. Better Chicken Commitment, EU Broiler Chicken Welfare: A Comprehensive Guide to the Better Chicken Commitment (2023), https://betterchickencommitment.com/eu-broiler-chicken-welfare.pdf; Ingrid C. Jong, 'Welfare Issues in Poultry Housing and Management: Broilers,' in Understanding the Behaviour and Improving the Welfare of Chickens, eds. Joy A. Mench and Michael C. Appleby (Wageningen Academic Publishers, 2020), https://edepot.wur.nl/539487.

165. Karin Björk, 'Sveriges omställning till burfria system för värphöns: En studie om hur aktörer inom äggbranschen upplever processen' [Sweden's Transition to

Cage-Free Systems for Laying Hens: A Study on How Actors in the Egg Industry Experience the Process], Master's Thesis, Swedish University of Agricultural Sciences, 2013), https://stud.epsilon.slu.se/5609/1/Bjork_K_130524.pdf.

166. Farm Animal Welfare Council, FAWC Report on the Welfare of Broiler Breeders (Department for Environment, Food and Rural Affairs, 2012), https://assets. publishing.service.gov.uk/government/uploads/system/uploads/attachment_data/file/325543/FAWC_report_on_the_welfare_of_broiler_breeders.pdf; House of Commons, Environment, Food and Rural Affairs Committee, Poultry Farming in the United Kingdom: Thirteenth Report of Session 2002–03, Vol. 1, HC 779-I (The Stationery Office, 2003), https://publications.parliament.uk/pa/cm200203/cmselect/cmenvfru/779/779.pdf.

167. Hubbard Breeders, 'Specific Labels and Markets,', https://www.hubbardbreeders.com/premium/specific-labels-markets/; Neil Dullaghan, Case Studies of Farmed Animal Legislation in the European Union (Rethink Priorities, September 21, 2021), https://docs.google.com/document/d/1zZW4ae9Cw7LdsS--m2c81_PZPy9Gye95mtHE2lKt1HA/.

168. European Commission, Scientific Committee on Animal Health and Animal Welfare, The Welfare of Chickens Kept for Meat Production (Broilers), SANCO.B.3/AH/R15/2000 (European Commission, 2000), https://food.ec.europa.eu/system/files/2020-12/sci-com_scah_out39_en.pdf.

169. 'RSPCA Urges End to Chicken Cruelty,' BBC News, November 23, 2001, http://news.bbc.co.uk/2/hi/uk_news/1671690.stm.

170. Compassion in World Farming Limited v. The Secretary of State for the Environment, Food and Rural Affairs, Case No. CO/1779/2003, filed 2003, available at Michigan State Univeresity, Animal Legal & Historical Center, https://www.animallaw.info/pleading/compassion-world-farming-limited-vthe-secretary-state-environment-food-and-rural-affairs; Jacky Turner et al., The Welfare of Broiler Chickens in the European Union (CIWFompassion in World Farming Trust, 2005), https://www.ciwf.org.uk/media/3818904/welfare-of-broilers-in-the-eu.pdf.

171. Select Committee on Environment, Food and Rural Affairs, Minutes of Evidence, Memorandum submitted by the British Poultry Council (x19), prepared July 23, 2003, https://publications.parliament.uk/pa/cm200203/cmselect/cmenvfru/779/3061708.htm.

172. Council of the European Union, Annex to the Proposal for a Council Directive laying down minimum rules for the protection of chickens kept for meat production, Impact Assessment, ST 9606/05 ADD 1, June 13, 2005, https://data.consilium.europa.eu/doc/document/ST-9606-2005-ADD-1/en/pdf.

173. Council of the European Union, Presidency of the Council of the EU, Full list of the Presidencies of the Council of the EU, DG Communication, https://www.consilium.europa.eu/media/rgpdgmgc/past-presidencies.pdf.

174. European Commission, 'Commission Proposes Legislation to Improve Welfare of Broiler Chickens,' press release IP/05/637, May 31, 2005, https://ec.europa.eu/commission/presscorner/detail/en/ip_05_637; Global Ag Media, 'The EC Proposes the Legislation for Its New Broiler Welfare Directive,' Poultry Site, June 21, 2005, https://www.thepoultrysite.com/news/2005/06/the-ec-proposes-the-legislation-for-its-new-broiler-welfare-directive.

175. European Commission, Proposal for a Council Directive laying down minimum rules for the protection of chickens kept for meat production, COM(2005) 221 final, Brussels: Commission of the European Communities, May 30, 2005, https://eur-lex.europa.eu/legal-content/EN/TXT/PDF/?uri=CELEX:52005PC0221.

176. European Commission, 'Commission Proposes Legislation to Improve Welfare of Broiler Chickens';
Commission of the European Communities, Commission staff working document: Annex to the proposal for a Council Directive laying down minimum rules for the protection of chickens kept for meat production — Impact assessment' (SEC(2005) 801 final), June 10, 2005, https://eur-lex.europa.eu/legal-content/EN/TXT/HTML/?uri=CELEX:52005SC0801.

177. Thomson et al., 'New Dataset on Decision-Making.'

178. Council of the European Union, '2676th Council Meeting – Agriculture and Fisheries, Brussels, 18 July 2005,' 10817/05, press release, July 18, 2005, https://ec.europa.eu/commission/presscorner/api/files/document/print/en/pres_05_179/PRES_05_179_EN.pdf;
Council of the European Union, Result of Procedure: Proposal for a Council directive laying down minimum rules for the protection of chickens kept for meat production – Expected socio-economic impact on operators and consumers, ST 7522/06, March 21, 2006, https://data.consilium.europa.eu/doc/document/ST-7522-2006-INIT/en/pdf.

179. European Economic and Social Committee, Opinion on the Proposal for a Council Directive for a Council Directive laying down minimum rules for the protection of chickens kept for meat production (COM[(2005)] 221 final – — 2005/0099 CNS).

180. European Parliament, Committee on Agriculture and Rural Development, Report on the Proposal for a Council Directive Laying Down Minimum Rules for the Protection of Chickens Kept for Meat Production (Rapporteur: Thijs Berman), A6-0017/2006 (February 1, 2006), https://www.europarl.europa.eu/doceo/document/A-6-2006-0017_EN.pdf.

181. Council of the European Union, Result of Procedure, ST 7522/06;
Sveriges Riksdag, EU-nämnden, Appendix to Document from the EU Committee 2005/06:1375, Top. 1, DP 09 Chicken 7522 06, 2006, https://www.riksdagen.se/sv/dokument-och-lagar/dokument/bilaga-till-dokument-fran-eu-namnden/dp-09-kyckling-7522-06_gt0n1375/.

182. Council of the European Union, Proposal for a Council Directive laying down minimum rules for the protection of chickens kept for meat production (COM(2005) 221 final) – Outcome of the European Parliament's first reading, ST 16263/06, December 12, 2006, https://data.consilium.europa.eu/doc/document/ST-16263-2006-INIT/en/pdf.

183. Poultry World Reporters, 'EU Welfare Standards Different,' Farmers Weekly, January 11, 2007, https://www.fwi.co.uk/business/eu-welfare-standards-different.

184. Council of the European Union, 2797th Council Meeting – Agriculture and Fisheries, Brussels, 7 May 2007, 8426/07, press release, May 7, 2007, https://www.minagric.gr/images/stories/agropol/En/Agro_pol/COUNCIL/94008.pdf;
Council of the European Union, Council Directive 2007/43/EC of 28 June 2007 laying down minimum rules for the protection of chickens kept for meat pro-

duction, Official Journal of the European Union L 182, July 12, 2007, 19–28, https://eur-lex.europa.eu/eli/dir/2007/43/oj/eng.

185. Council of the European Union, Addendum to the Proposal for a Council laying down minimum rules for the protection of chickens kept for meat production – Expected socio-economic impact on economic operators and consumers,ST 10308/07 ADD 1, June 13, 2007, https://data.consilium.europa.eu/doc/document/ST-10308-2007-ADD-1/en/pdf.

186. McKenna, The Case Against the Veal Crate.

187. McKenna, The Case Against the Veal Crate.

CHAPTER 4: MAKING PROGRESS STICK: HOW THE LAWS WERE ENFORCED

188. See nn. 1–8.

189. European Commission, Single Market Infringement Cases, Single Market and Competitiveness Scoreboard, reporting period December 2023–November 2024, https://single-market-scoreboard.ec.europa.eu/enforcement-tools/infringements_en.

190. European Court of Auditors (ECA), Special Report No. 31/2018: Animal Welfare in the EU: Closing the Gap Between Ambitious Goals and Practical Implementation (November 13, 2018), https://www.eca.europa.eu/en/publications?did=47557;
ECA, Special Report No 26/2016: Making Cross-Compliance More Effective and Achieving Simplification Remains Challenging, (October 27, 2016), https://www.eca.europa.eu/en/publications?did=38185.

191. Dimiter Toshkov, 'Compliance with EU Law in Central and Eastern Europe: The Disaster That Didn't Happen (Yet),' L'Europe en Formation 364, no. 2 (2012): 91–109, https://doi.org/10.3917/eufor.364.0091;
Enrico Borghetto and Fabio Franchino, 'The Role of Subnational Authorities in the Implementation of EU Directives,' Journal of European Public Policy 17, no. 6 (2010): 759–80, https://doi.org/10.1080/13501763.2010.486972;
Miriam Hartlapp and Gerda Falkner, 'Problems of Operationalization and Data in EU Compliance Research,' European Union Politics 10, no. 2 (2009): 281–304, https://doi.org/10.1177/1465116509103370;
Thomas König and Brooke Luetgert, 'Troubles with Transposition? Explaining Trends in Member-State Notification and the Delayed Transposition of EU Directives,' British Journal of Political Science 39, no. 1 (2009): 163–94, https://doi.org/10.1017/S0007123408000380;
Gerda Falkner and Oliver Treib, 'Three Worlds of Compliance or Four? The EU-15 Compared to New Member States,' JCMS Journal of Common Market Studies 46, no. 2 (2008): 293–313, https://doi.org/10.1111/j.1468-5965.2007.00777.x;
Michael Kaeding, 'In Search of Better Quality of EU Regulations for Prompt Transposition: The Brussels Perspective,' European Law Journal 14, no. 5 (2008): 583–603, https://doi.org/10.1111/j.1468-0386.2008.00431.x;
Gerda Falkner et al., 'Worlds of Compliance: Why Leading Approaches to European Union Implementation Are Only "Sometimes-True Theories,"' European Journal of Political Research 46, no. 3 (2007): 395–416, https://doi.org/10.1111/j.1475-6765.2007.00703.x;

Markus Haverland and Marleen Romeijn, 'Do Member States Make European Policies Work? Analysing the EU Transposition Deficit,' Public Administration 85, no. 3 (2007): 757–78, https://doi.org/10.1111/j.1467-9299.2007.00670.x;
Robert Thomson et al., 'The Paradox of Compliance: Infringements and Delays in Transposing European Union Directives,' British Journal of Political Science 37, no. 4 (2007): 685–709, https://doi.org/10.1017/S0007123407000373;
Ellen Mastenbroek, 'EU Compliance: Still a "Black Hole"?,' Journal of European Public Policy 12, no. 6 (2005): 1103–20, https://doi.org/10.1080/13501760500270869;
Ulf Sverdrup, 'Compliance and Conflict Management in the European Union: Nordic Exceptionalism,' Scandinavian Political Studies 27, no. 1 (2004): 23–43, https://doi.org/10.1111/j.1467-9477.2004.00098.x;
Christoph Knill and Andrea Lenschow, 'Coping With Europe: The Impact of British and German Administrations on the Implementation of EU Environmental Policy,' Journal of European Public Policy 5, no. 4 (1998): 595–614, https://doi.org/10.1080/13501769880000041.

192. Tanja A. Börzel et al., 'Obstinate and Inefficient: Why Member States Do Not Comply With European Law,' Comparative Political Studies 43, no. 11 (2010): 1363–90, https://doi.org/10.1177/0010414010376910;
Falkner and Treib, 'Three Worlds of Compliance or Four?';
Falkner et al., 'Worlds of Compliance.'

193. König and Luetgert, 'Troubles With Transposition?'

194. Hartlapp and Falkner, 'Problems of Operationalization';
Kaeding, 'In Search of Better Quality';
Haverland and Romeijn, 'Do Member States Make European Policies Work?';
Thomson et al., 'Paradox of Compliance';
Mastenbroek, 'EU Compliance.'

195. Council of the European Union, Working Party on Animals and Veterinary Questions, https://www.consilium.europa.eu/en/council-eu/preparatory-bodies/working-party-on-animals-and-veterinary-questions/.

196. Council of the European Union, Animal Welfare – An Integral Part of Sustainable Animal Production, Outcome of the Presidency Questionnaire, ST 6007/20, February 20, 2020, https://data.consilium.europa.eu/doc/document/ST-6007-2020-INIT/en/pdf.

197. Council of the European Union, Animal Welfare – An Integral Part of Sustainable Animal Production, Outcome of the Presidency Questionnaire, ST 6007/20;
EU Platform on Animal Welfare, 'EU Animal Welfare Strategy 2012–2015: Achievements and Future Outlook' paper presented at the 10th Meeting of the EU Platform on Animal Welfare, Brussels, October 7, 2019, https://food.ec.europa.eu/document/download/e76e92b6-82d4-4a4a-b244-27b11ffd0863_en?-filename=aw_platform_20191007_pres-01.pdf.

198. Council of the European Union, Council Directive 2008/120/EC of 18 December 2008 laying down minimum standards for the protection of pigs, Official Journal of the European Union L 47, February 18, 2009, 5–13, https://eur-lex.europa.eu/legal-content/EN/TXT/PDF/?uri=CELEX:32008L0120.

199. Dale A. Sandercock et al., 'Transcriptomics Analysis of Porcine Caudal Dorsal Root Ganglia in Tail Amputated Pigs Shows Long-Term Effects on Many

Pain-Associated Genes,' Frontiers in Veterinary Science 6 (2019), https://doi.org/10.3389/fvets.2019.00314.

200. Irene Camerlink and Winanda W. Ursinus, 'Tail Postures and Tail Motion in Pigs: A Review,' Applied Animal Behaviour Science 230 (2020): 105079, https://doi.org/10.1016/j.applanim.2020.105079.

201. European Food Safety Authority, 'Scientific Opinion of the Panel on Animal Health and Welfare on the Risks Associated with Tail Biting in Pigs and Possible Means to Reduce the Need for Tail Docking Considering the Different Housing and Husbandry Systems,' EFSA Journal 5, no. 12 (2007): 1–13, https://doi.org/10.2903/j.efsa.2007.611;

202. Nancy De Briyne et al., 'Phasing Out Pig Tail Docking in the EU: Present State, Challenges and Possibilities,' Porcine Health Management 4, no. 1 (2018): 27. https://doi.org/10.1186/s40813-018-0103-8

203. De Briyne et al., 'Phasing Out Pig Tail Docking in the EU.'

204. Council of the European Union, Council Directive 2008/120/EC of 18 December 2008 laying down minimum standards for the protection of pigs, Official Journal of the European Union L 47, February 18, 2009, 5–13, https://eur-lex.europa.eu/legal-content/EN/TXT/PDF/?uri=CELEX:32008L0120.

205. De Briyne et al., 'Phasing Out Pig Tail Docking in the EU.' Enrichment materials numbers come from this source, but the paper omitted a number of countries.

206. Council of the European Union, Council Directive 98/58/EC of 20 July 1998 concerning the protection of animals kept for farming purposes, Official Journal of the European Communities L 221, August 8, 1998, 23–27, https://eur-lex.europa.eu/legal-content/EN/TXT/?uri=celex%3A31998L0058.

207. Alice Di Concetto et al., The Treatment of Farmed Fish Under EU Law, Research Note 7 (The European Institute for Animal Law & Policy, 2024), https://animallaweurope.org/wp-content/uploads/Research-Note-7_The-Treatment-of-Farmed-Fish-Under-EU-Law.pdf; European Commission, Commission Staff Working Document: Evaluation of the EU Animal Welfare Legislation, SWD(2022) 328 final, October 4, 2022, https://food.ec.europa.eu/system/files/2022-10/aw_eval_revision_swd_2022-328_en.pdf.

208. European Commission, 'Animal Welfare: Commission Report Confirms the Potential Benefits of Banning Conventional Battery Cages for Laying Hens,' press release, January 8, 2008, https://ec.europa.eu/commission/presscorner/detail/en/ip_08_19.

209. Council of the European Union, Council Directive 1999/74/EC of 19 July 1999 laying down minimum standards for the protection of laying hens, Official Journal of the European Communities L 203, August 3, 1999, 53–57, https://eur-lex.europa.eu/eli/dir/1999/74/oj/eng.

210. Agra CEAS Consulting Ltd., Study on the Socio-Economic Implications of the Various Systems to Keep Laying Hens, final report for the European Commission, (December 2004), https://www.yumpu.com/en/document/view/42854993/2120-final-reportpdf-agra-ceas-consulting;
Taina Vesanto et al., Työryhmämuistio MMM 2003:14, Kananmunien tuotantostrategia: Uusiin tuotantomuotoihin siirtyminen vuonna 2012, Loppuraportti [Working Group Memorandum MMM 2003:14, Chicken Egg Production Strategy: Transition to New Forms of Production in 2012, Final Report]

(Maa- ja metsätalousministeriö , 2003), https://julkaisut.valtioneuvosto.fi/
bitstream/handle/10024/160545/trm%25202003_14_Kananmunien%2520tu-
otantostrategia_Uusiin%2520tuotantomuotoihin%2520siirtyminen%2520vuon-
na%25202012.%2520Loppuraportti.pdf.

211. European Commission, 'Animal Welfare: Commission Report.'

212. Farm Animal Welfare Council, FAWC Opinion on Enriched Cages for Laying
Hens (Department for Environment, Food & Rural Affairs, November 15, 2007),
https://www.gov.uk/government/publications/fawc-opinion-on-enriched-cag-
es-for-laying-hens.

213. Agra CEAS Consulting Ltd., Study on the Socio-Economic Implications of the
Various Systems to Keep Laying Hens.

214. Daniel Wild, 'Moves Begin to Block the Cage Ban,' FarmingUK, September
25, 2005, https://www.farminguk.com/news/moves-begin-to-block-the-cage-
ban_231.html;
Natalie Kinsley, 'Poland Calls on Delaying Cage Ban to 2017,' Poultry World,
February 17, 2010, https://www.poultryworld.net/poultry/poland-calls-on-de-
laying-cage-ban-to-2017/.

215. European Parliament, European Parliament Resolution of 16 December 2010
on the EU laying hens industry: the ban on the use of battery cages from
2012, EUR-Lex 52010IP0493, https://eur-lex.europa.eu/legal-content/EN/TX-
T/?uri=CELEX%3A52010IP0493.

216. European Parliament, European Parliament Resolution of 16 December 2010 on
the EU Laying Hens Industry states: 'whereas in 2008 the Commission adopted
a communication on the various systems of rearing laying hens, in particular
those covered by Council Directive 1999/74/EC, confirming that the decision to
phase out battery cages by 1 January 2012 was justified and that no amendment
to the directive was necessary; whereas the Commission reiterated this position
at the Agriculture Council meeting on 22 February 2010. '

217. House of Commons Environment, Food and Rural Affairs Committee, The
Welfare of Laying Hens Directive—Implications for the Egg Industry, Ninth
Report of Session 2010–12, HC 830 (The Stationery Office Limited, September
2, 2011), https://www.farminguk.com/content/knowledge/efra%20report.pdf.

218. The exact figure is unclear. In mid-2010 it was forecast that about 30% of hens
would still be in conventional battery cages by 2012 (Nick Chippindale, 'The
2012 EU Ban on Conventional Cages and Its Effect,' Nuffield Farming Schol-
arships Trust, December 2010, https://www.nuffieldscholar.org/sites/default/
files/reports/2009_UK_Nick-Chippindale_The-2012-Eu-Ban-On-Convention-
al-Cages-And-Its-Effect.pdf). Data from April 2011 showed approximately 144
million hens, more than a third of the EU total, were still being kept in conventional battery cages (Francesco Guarascio and Charlie Dunmore, 'EU Countries Face Legal Action over Hen Cage Ban,' Reuters, January 23, 2012, https://
www.reuters.com/article/us-eu-hens-welfare-idUSTRE80M1IY20120123/). A
European Commission spokesman reported in 2012 that 46 million hens remained in conventional battery cages, about 14% of the total number of birds
('UK Battery Farms Break EU Rules,' BBC News, January 13, 2012, https://www.
bbc.com/news/uk-16540769). However, other sources reported higher figures of
83-84 million hens (for example, Stuart Agnew, 'An Instant Ban on Europe's
Battery Eggs Won't Work,' Guardian, September 5, 2011, https://www.theguard-

ian.com/commentisfree/2011/sep/05/british-egg-farmers-battery-ban), while Hans-Wilhelm Windhorst suggested 43.5 million hens in June 2012 (Hans-Wilhelm Windhorst, 'The European Egg Industry in Transition,' paper presented at the IEC Berlin Conference, September 23, 2015, slide 22, https://slideplayer.com/slide/12745069/).

219. Between January 2011, when over 364 million layers were kept, and January 2012, when the transformation process reached a critical phase, the number of layers decreased by almost 33 million birds (Mark Clements, 'Lessons to Be Learned From European Union's Cage Ban,' WATTPoultry, May 20, 2014, https://www.wattagnet.com/egg/egg-production/article/15512998/lessons-to-be-learned-from-european-unions-cage-ban). Some countries were quite persistent in their efforts to find a way out, continuing to allow new conventional battery cages in 2011, even though this was supposed to have stopped in 2003. France and Greece were highlighted as doing this in DG SANTE reports.

220. One report said 75% of the EU's sows were on compliant holdings (WATTAgNet Staff, '17 EU Countries Are Not Sow Stall Compliant,' WATTPoultry, December 20, 2012, https://www.wattagnet.com/home/article/15505939/17-eu-countries-are-not-sow-stall-compliant); others suggested 2 million pigs per week or 40% of EU pig production will not be compliant (David Burch, 'EU Gestation Stalls Ban,' Pig Progress, January 4, 2013, https://www.pigprogress.net/pigs/eu-gestation-stalls-ban/ and Daniel Wild, 'Europe's Pig Farmers Flout Brussels Stalls Ban,' FarmingUK, December 31, 2012, https://www.farminguk.com/news/europe-s-pig-farmers-flout-brussels-stalls-ban_24753.html). Neither report mentions pregnant/breeder pigs specifically. Based on 2012 compliance numbers and Eurostat breeding sow numbers, the actual number of pigs was 4,093,364.

221. European Commission, Commission Staff Working Paper: Impact Assessment, SEC(2012) 55 final, January 19, 2012, https://eur-lex.europa.eu/legal-content/EN/TXT/PDF/?uri=CELEX:52012SC0055.

222. Helen Lambert, No Animal Left Behind: Phasing Out Cages in the EU; The Road to a Smooth Transition (Eurogroup for Animals, March 2023), https://www.eurogroupforanimals.org/files/eurogroupforanimals/2023-03/NALB-Phasing%20out%20cages-final.pdf.

223. European Commission, DG Health and Food Safety, Welfare of Cattle on Dairy Farms (Publications Office of the European Union, 2017), https://op.europa.eu/en/publication-detail/-/publication/8950fa88-d651-11e7-a506-01aa75ed71a1/language-en.

224. European Commission: Directorate-General for Health and Food Safety, Welfare of Cattle on Dairy Farms : Overview Report (Publications Office of the European Union, 2017), https://op.europa.eu/en/publication-detail/-/publication/8950fa88-d651-11e7-a506-01aa75ed71a1/language-en.

225. Bettina B. Bock et al., Report on the Main Problem Areas and Their Sensitivity to Be Addressed by Knowledge Transfer for Each of the Specific Aspects of the Legislation Chosen for This Project' (EuWelNet, April 7, 2014), https://www.euwelnet.eu/media/1158/euwelnet_deliverable_4_final.pdf.

226. European Commission, Directorate-General for Health and Food Safety, Directorate for Health and Food Audits and Analysis, Use of Slaughterhouse Data to Monitor Welfare of Broilers on Farm – Overview Report (Publications Office of the European Union, 2016), 6, https://data.europa.eu/doi/10.2772/57892.

227. European Commission: Directorate-General for Health and Food Safety, Study on the Application of the Broiler Directive DIR 2007/43/EC and Development of Welfare Indicators: Final Report, (Publications Office, 2017) https://op.europa.eu/en/publication-detail/-/publication/f4ccd35e-d004-11e7-a7df-01aa75e-d71a1/language-en.

228. Pig333, 'European Union: Sow Stall Update,' February 7, 2013, https://www.pig333.com/latest_swine_news/european-union-sow-stall-update_6825/.

229. 3trois3, 'Europe: Niveau d'application de la loi sur le bien-être' [Europe: Level of Application of Welfare Law], January 8, 2013, https://www.3trois3.com/articles/europe-niveau-dapplication-de-la-loi-sur-le-bien-etre_10216/; WATTPoultry, '17 EU Countries Are Not Sow Stall Compliant,' January 15, 2013, https://www.wattagnet.com/home/article/15505939/17-eu-countries-are-not-sow-stall-compliant; WATTPoultry, '17 EU Countries Are Not Sow Stall Compliant,' January 15, 2013, https://www.wattagnet.com/home/article/15505939/17-eu-countries-are-not-sow-stall-compliant.

230. European Commission, Enforcement Tools, The Single Market and Competitiveness Scoreboard, https://single-market-scoreboard.ec.europa.eu/enforcement-tools_en.

231. European Commission, Single Market Infringement Cases, The Single Market and Competitiveness Scoreboard, https://single-market-scoreboard.ec.europa.eu/enforcement-tools/infringements_en.

232. Philip Clark, 'Video: Illegal Egg Farm Still Operating in Greece,' Farmers Weekly, July 10, 2014, https://www.fwi.co.uk/livestock/poultry/video-illegal-egg-farm-still-operating-in-greece. In 2013, the Commission brought Italy and Greece to court, where around 20 million hens still remained in conventional battery cages. Both countries were found guilty in 2014 but in the end they only had to pay legal costs, not financial penalties, possibly due to the financial crisis. The cases against them were not formally closed until October-November 2015, suggesting battery cages may have remained in use in both countries for almost four years after the deadline. See CIWF, 'Greece Gets Away with It,' September 5, 2014, https://www.ciwf.org.uk/news/greece-gets-away-with-it/; WATTPoultry, 'Italy Found Guilty of Non-Compliance with Cage Ban,' May 30, 2014, https://www.wattagnet.com/broilers-turkeys/article/15510040/italy-found-guilty-of-non-compliance-with-cage-ban.

233. CIWF, 'Six Countries Still Flouting Sow Stall Ban,' January 29, 2014, https://www.ciwf.org.uk/news/six-countries-still-flouting-sow-stall-ban/.

234. Council of the European Union, Animal Welfare – An Integral Part of Sustainable Animal Production, Outcome of the Presidency Questionnaire, ST 6007/20.

235. For cattle registration example, see European Court of Auditors, Special Report No. 26/2016: Making Cross-Compliance More Effective and Achieving Simplification Remains Challenging (Publications Office of the European Union, 2016), 30, https://www.eca.europa.eu/Lists/ECADocuments/SR16_26/SR_CROSS_COMPLIANCE_EN.pdf; European Commission, Food and Veterinary Office, General Report of Findings from Missions 2008-2010 on Enforcement of Animal Welfare Standards for Laying Hens Kept in Unenriched Cages (European Commission, 2011), archived April 30, 2014, https://web.archive.org/web/20140430021629/http://ec.europa.eu/food/fvo/specialreports/2011_8814_en.pdf.

236. European Commission, Food and Veterinary Office, General Report of Findings from Missions 2008-2010.

237. European Commission, DG Health and Food Safety, Audit Reports: Health and Food Audits and Analysis, https://ec.europa.eu/food/audits-analysis/audit_reports/index.cfm.

238. European Commission, Food and Veterinary Office, General Report of Findings from Missions 2008-2010.

239. Jill Fernandes et al., 'Addressing Animal Welfare Through Collaborative Stakeholder Networks,' Agriculture 9, no. 6 (2019): 132, https://doi.org/10.3390/agriculture9060132.

240. European Commission, Special Eurobarometer 442: Attitudes of Europeans towards Animal Welfare (March 2016), https://europa.eu/eurobarometer/surveys/detail/2096.

241. Council of the European Union, Outcome of the Presidency Questionnaire: Animal Welfare, ST 6007/20.

242. European Commission, 20th Annual Report on Monitoring the Application of Community Law (2002), COM(2003) 669 final, November 21, 2003, https://eur-lex.europa.eu/legal-content/EN/TXT/PDF/?uri=CELEX:52003DC0669.

243. European Parliament, Answer to Parliamentary Question E-012150/13, November 26, 2013, https://www.europarl.europa.eu/doceo/document/E-7-2013-012150-ASW_EN.html;
European Parliament, Parliamentary Question: Failure by Spain to Enforce the Directive on Pig Welfare (Written Question E-012150/2013), October 23, 2013, https://www.europarl.europa.eu/doceo/document/E-7-2013-012150_EN.html.

244. CIWF, 'Greek Government Condemned for Failure to Enforce Animal Welfare Laws,' September 14, 2009, https://www.ciwf.org.uk/news/greek-government-condemned-for-failure-to-enforce-animal-welfare-laws/.

245. European Parliament, Draft Agenda: Thursday, 3 October 2024, PETI_OJ(2024)10-03_1, October 3, 2024, https://www.europarl.europa.eu/doceo/document/PETI-OJ-2024-10-03-1_EN.html.

CHAPTER 5: A EUROPEAN LOVE AFFAIR WITH CAGE FREE

246. EU Reporter Correspondent, '#EndTheCageAge: NGOs, MEPs and EU Citizens Unite to Celebrate Success of the European Citizens' Initiative (ECI),' Eureporter, October 8, 2019, https://www.eureporter.co/frontpage/2019/10/08/endthecageage-ngos-meps-and-eu-citizens-unite-to-celebrate-success-of-the-european-citizens-initiative-eci/;
European Commission, End the Cage Age, European Citizens' Initiative, https://citizens-initiative.europa.eu/initiatives/details/2018/000004/end-cage-age_en.

247. Council of the European Union, Council Directive 1999/74/EC of 19 July 1999 laying down minimum standards for the protection of laying hens, Official Journal of the European Communities L 203, August 3, 1999, https://eur-lex.europa.eu/legal-content/EN/TXT/PDF/?uri=CELEX:31999L0074.

248. Konrad Lorenz, 'Tiere sind Gefühlsmenschen' [Animals Are Emotional Beings], Der Spiegel, November 16, 1980, https://www.spiegel.de/wissenschaft/tiere-sind-gefuehlsmenschen-a-20aa3d2f-0002-0001-0000-000014329375;

'Konrad Lorenz: Facts,' Nobel Prize, https://www.nobelprize.org/prizes/medicine/1973/lorenz/facts/.

249. Agico Group, 'Aviary System for Laying Hens,', https://chickenlayercage.com/cage-free-chicken-house/aviary-system-for-laying-hens/.

250. Welfare Footprint Institute, 'The Welfare of Laying Hens: Quantifying the Welfare Impact of the Transition to Indoor Cage-Free Housing Systems,', https://welfarefootprint.org/laying-hens/.

251. RSPCA, Indoor Barn: Understanding Verandas for Laying Hens, https://business.rspcaassured.org.uk/media/pwfd3i1w/indoor-barn_-understanding-verandas-for-laying-hens.pdf.

252. Austin Alonzo, 'Will the Cage-Free Housing of Choice Be Combi Systems?' WATTPoultry, July 13, 2016, https://www.wattagnet.com/egg/egg-production/article/15518313/will-the-cage-free-housing-of-choice-be-combi-systems-wattagnet.

253. European Commission, 'Eggs – Market Situation – Dashboard.'

254. Rashi Singh, 'Egg Market Outlook: 2019 and Beyond,' The Smart Cube, February 5, 2019, https://www.thesmartcube.com/resources/blog/egg-market-outlook-2019-and-beyond/.

255. Mose Apelblat, 'Commissioner with Double Portfolio Confirms Focus on Health, Vague on Animal Welfare,' Brussels Times, November 8, 2024, https://www.brusselstimes.com/1303917/commissioner-with-double-portfolio-confirms-focus-on-health-vague-on-animal-welfare.

256. Dietmar Hipp and Renate Nimtz-Köster, 'Sündenfall der Menschheit' [Humanity's Original Sin], Der Spiegel, July 4, 1999, https://www.spiegel.de/wissenschaft/suendenfall-der-menschheit-a-81ee49bf-0002-0001-0000-000013918400.

257. Federal Constitutional Court of Germany, Urteil des Zweiten Senats [Judgment of the Second Senate],2 BvF 3/90, 6 July 1999, https://www.bundesverfassungsgericht.de/SharedDocs/Entscheidungen/DE/1999/07/fs19990706_2bvf000390.html.

258. European Parliament, Answer to Written Question E-4011/06, October 31, https://www.europarl.europa.eu/doceo/document/E-6-2006-4011-ASW_DA.html.

259. Tierschutzgesetz [Animal Welfare Act] (Germany), as amended August 4, 2006, https://www.buzer.de/gesetz/4804/v144110-2006-08-04.htm.

260. Neil Dullaghan, 'Laying Hens by Way of Keeping, All Years (DG AGRI data),' EU Animal Welfare Law Compliance Data, Google Sheets,

261. European Commission, Report from the Commission to the Council with Regard to Developments in Consumption, Washing and Marking of Eggs, COM(2003) 479 final, August 12, 2003, https://eur-lex.europa.eu/legal-content/EN/TXT/HTML/?uri=CELEX%3A52003DC0479.

262. RSPCA Australia, 'Eggcellent News: ALDI Makes Cage Free Commitment,' May 25, 2016, https://www.rspca.org.au/latest-news/media-centre/eggcellent-news-aldi-makes-cage-free-commitmen.

263. Ute Knierim et al., Alternative Legehennenhaltung in der Praxis: Erfahrungen, Probleme, Lösungsansätze [Alternative Laying Hen Husbandry in Practice: Experiences, Problems, Solutions], Thünen Report 75 (Thünen-Institut, 2019), https://literatur.thuenen.de/digbib_extern/zi041750.pdf.

264. Henrik Ballwanz, 'Streit um Käfighaltung' [Dispute Over Caged Animals], Deutschlandfunk, October 8, 2004, https://www.deutschlandfunk.de/streit-um-kaefighaltung-100.html.

265. Tierschutz-Nutztierhaltungsverordnung [Animal Welfare – Livestock Ordinance], as amended April 22, 2016, https://www.buzer.de/gesetz/7344/v198072-2016-04-22.htm.

266. PROVIEH, 'Raus aus dem Käfig!' [Out of the Cage!], Provieh Magazine, no. 1 (2011), https://www.provieh.de/wp-content/uploads/2021/01/provieh_magazin_2011_01.pdf.

267. Dullaghan, 'Laying Hens by Way of Keeping, All Years.'

268. Bundesrat, Entwurf einer Fünften Verordnung zur Änderung der Tierschutz-Nutztierhaltungsverordnung [Draft of a Fifth Ordinance Amending the Animal Welfare Livestock Farming Ordinance], Drucksache 95/12 (Beschluss), March 2, 2012, https://umwelt-online.de/PDFBR/2012/0095_2D12B.pdf; 'Bundesverfassungsgericht rügt Legehennen-Verordnung [Federal Constitutional Court criticizes laying hen ordinance],' Stimme.de, December 2, 2010, https://www.stimme.de/politik/welt/politik/dw/bundesverfassungsgericht-ruegt-legehennen-verordnung-art-2001830.

269. Foodwatch e.V., Warum kaufen Sie eigentlich so viele Käfigeier? Ein Hintergrundpapier von Foodwatch zum Lebensmittel Ei [Why Do You Buy So Many Cage Eggs? A Background Paper from Foodwatch on the Foodstuff Egg], updated April 2012, https://www.foodwatch.org/fileadmin/_migrated/content_uploads/hintergrund_eier_kaefigeier_20120400.pdf.

270. Landesvertretung Rheinland-Pfalz, Entwurf einer Fünften Verordnung zur Änderung der Tierschutz-Nutztierhaltungsverordnung (Legehennen-Verordnung) [Draft of a Fifth Ordinance Amending the Animal Welfare Livestock Farming Ordinance (Laying Hen Ordinance)], Bundesrat Drucksache 95/12, March 2, 2012, https://landesvertretung.rlp.de/im-bund/mitwirken-im-bund/detail/entwurf-einer-fuenften-verordnung-zur-aenderung-der-tierschutz-nutztierhaltungsverordnung-legehennen-verordnung;
Ministerium für Klimaschutz, Umwelt, Energie und Mobilität Rheinland-Pfalz, 'Kleingruppenhaltung bei Legehennen: Rheinland-Pfalz und Niedersachsen schlagen Übergangsfrist bis 2023 vor [Small Group Housing for Laying Hens: Rhineland-Palatinate and Lower Saxony Propose Transitional Period until 2023],' press release, February 25, 2021, https://mkuem.rlp.de/service/pressemitteilungen/detail/kleingruppenhaltung-bei-legehennen-rheinland-pfalz-und-niedersachsen-schlagen-uebergangsfrist-bis-2023-vor;
Deutscher Tierschutzbund e.V., 'Das kurze Hühnerleben in der Landwirtschaft [The Short Life of Chickens in Agriculture],', https://www.tierschutzbund.de/tiere-themen/tiere-in-der-landwirtschaft/huehner/#c7467.

271. Tierschutz-Nutztierhaltungsverordnung [Animal Welfare – Livestock Ordinance].

272. VGT, Animal Law: What VGT Has Achieved in Austria!, December 11, 2007, https://www.abolitionistapproach.com/media/links/p140/another-portion.pdf.

273. VGT, Flächendeckende Gesetzesbrüche in heimischen Legebatterien,' [Widespread Legal Violations in Domestic Battery Farms],' August 4, 2003, https://vgt.at/de/aktuelles/detailseite/180/flaechendeckende-gesetzes-brueche-in-heimischen-legebatterien.html;

Tiroler Landesregierung, Gesetz vom 12. Dezember 1996 über die Raumord-
nung in Tirol (Tiroler Raumordnungsgesetz 1997–TROG 1997) [Law of December 12, 1996, on Spatial Planning in Tyrol (Tyrolean Spatial Planning Law 1997–
TROG 1(February 25, 1997), https://web.archive.org/web/20230521064515/
https://www.tirol.gv.at/fileadmin/_migrated/content_uploads/lgballe1997.pdf.

274. Commission of the European Communities, Report on the Welfare of Laying Hens
(Office for the Official Publications of the European Communities), November 3,
1998, https://aei.pitt.edu/10645/1/10645.pdf.

275. Martin Balluch, 'Austrian Battery Farm Campaign: First Signs of a Breakthrough,' United Poultry Concerns, May 17, 2004, https://www.upc-online.org/
battery_hens/51704austria.htm.

276. Martin Balluch, 'A Brief History of Austria's Current Campaign to Ban Battery
Cages,' United Poultry Concerns, May 18, 2004, https://www.upc-online.org/
battery_hens/51804austria.htm.

277. Martin Balluch, 'A Brief History of Austria's Current Campaign to Ban Battery Cages.'

278. Martin Balluch, 'A Brief History of Austria's Current Campaign to Ban Battery Cages.'

279. Martin Balluch, 'Legebatterien: Wenn uns die Vergangenheit einholt' [Battery Cages: When the Past Catches Up with Us], Martin Balluchs Blog, October 1, 2023,
https://martinballuch.com/legebatterien-wenn-uns-die-vergangenheit-einholt/;
Ian Traynor, 'Austria Bans Battery Cages,' Guardian, May 28, 2004, https://www.
theguardian.com/world/2004/may/28/animalwelfare.uk.

280. VGT, Großer Erfolg des VGT: Supermarktketten steigen aus Käfigeihandel aus'
[Great Success for VGT: Supermarket Chains Exit Cage Egg Trade], November
20, 2004, https://vgt.at/de/aktuelles/detailseite/1029/grosser-erfolg-des-vgt:-su-
permarktketten-steigen-aus-kaefigeihandel-aus.html.

281. Gyorgy Scrinis et al., 'The Caged Chicken or the Free-Range Egg? The Regulatory and Market Dynamics of Layer-Hen Welfare in the UK, Australia and
the USA,' Journal of Agricultural and Environmental Ethics 30, no. 6 (2017):
783–808, https://doi.org/10.1007/s10806-017-9699-y; Appleby, 'The EU Ban on
Battery Cages.'

282. SAFE For Animals, 'The Plight of Battery Hens has Received Recognition
Internationally,', https://safe.org.nz/our-work/animals-in-aotearoa/hens-2/
whats-happening-overseas/.

283. British Egg Industry Council, Written Evidence Submitted by the British Egg
Industry Council (EGG 14), in The Implications of the Welfare of Laying Hens
Directive for the Egg Industry: Written Evidence, (House of Commons Environment, Food and Rural Affairs Committee, February 9, 2011), https://pub-
lications.parliament.uk/pa/cm201011/cmselect/cmenvfru/writev/egg/egg.pdf.

284. Règlement grand-ducal du 28 janvier 2002 établissant les normes minimales
relatives à la protection des poules pondeuses [Grand-Ducal Regulation of January 28, 2002, Establishing Minimum Standards for the Protection of Laying
Hens], Journal Officiel, November 2, 2002, https://legilux.public.lu/eli/etat/leg/
rgd/2002/01/28/n1/jo.

285. Dullaghan, 'Laying Hens by Way of Keeping, All Years.'

286. Lindström, Hönan eller ägget.

287. Productivity Commission, Battery Eggs Sale and Production in the ACT (AusInfo, 1998), https://econwpa.ub.uni-muenchen.de/econ-wp/othr/papers/0107/0107009.pdf.

288. Ministry of Agriculture, Nature Management and Fisheries, Netherlands, Nota Dierenwelzijn [Policy Note on Animal Welfare], Parliamentary Document 28 286, no. 2 (Sdu Uitgevers, 2002), https://www.parlementairemonitor.nl/9353000/1/j9vvij5epmj1ey0/vi3akfpnl1zr#p1.

289. European Commission, Report from the Commission to the Council with Regard to Developments in Consumption, Washing and Marking of Eggs, COM(2003) 479 final.

290. Eurogroup for Animals, 'Calling on the Dutch Agricultural Minister to Ban Cages for Laying Hens,' January 16, 2019, https://www.eurogroupforanimals.org/news/calling-dutch-agricultural-minister-ban-cages-laying-hens.

291. J. Gautron et al., 'Review: What are the Challenges Facing the Table Egg Industry in the Next Decades and What Can be Done to Address Them?,' Animal 15, no. S1 (2021): 100282, https://doi.org/10.1016/j.animal.2021.100282.

292. Obránci zvířat, 'The Fight Is Over: The Ban on Cage Farming Has Been Signed by the President,' November 19, 2020, https://obrancizvirat.cz/podpis-prezidenta/.

293. OBRAZ – Obránci zvířat, Facebook, February 8, 2020, https://www.facebook.com/obrancizvirat/photos/a.900800736637427/2947921195258694.

294. Gabriela Brázdová, 'Češi vybojovali svobodu pro slepice. Zákaz klecových chovů podepsal prezident' [Czechs Won Freedom for Chickens: President Signed Ban on Battery Cages], Pozitivní zprávy, November 22, 2020, https://pozitivni-zpravy.cz/cesi-vybojovali-svobodu-pro-slepice-zakaz-klecovych-chodu-podepsal-prezident/.

295. Council of the European Union, Ban on Laying Hens in Cages: Information from the Czech Delegation, Document 10844/20, AGRI 273, FOOD 13, VETER 36 (September 16, 2020), https://data.consilium.europa.eu/doc/document/ST-10844-2020-INIT/en/pdf.

296. CIWF, 'The Slovak Republic to End the Cage Age for Hens,' February 14, 2020, https://www.ciwf.org.uk/news/the-slovak-republic-to-end-the-cage-age/.

297. TASR, 'Slovensko je sebestačné vo vajciach, otázne je dokedy' [Slovakia is Self-Sufficient in Eggs, the Question is How Long], Pravda, June 8, 2018, https://ekonomika.pravda.sk/ludia/clanok/472580-slovensko-je-sebestacne-vo-vajciach-otazne-je-dokedy/.

298. Eurogroup for Animals, 'An Ambitious Animal Welfare Code in Wallonia, Including a Ban on Battery Caged Hens,' April 27, 2018, https://www.eurogroupforanimals.org/news/ambitious-animal-welfare-code-wallonia-including-ban-battery-caged-hens.

299. Eurogroup for Animals, 'Denmark Bids Farewell to Cage Egg Production,' September 29, 2022, https://www.eurogroupforanimals.org/news/denmark-bids-farewell-cage-egg-production.

300. The Local Denmark, 'Denmark to Ban Caged Egg Production by 2035,' September 22, 2022, https://www.thelocal.dk/20220922/denmark-to-ban-caged-egg-production-by-2035.

301. Tony McDougal, 'France to Ban Sale of Eggs from Caged Hens by 2022,' Poultry World, February 21, 2018, https://www.poultryworld.net/poultry/france-to-ban-sale-of-eggs-from-caged-hens-by-2022/.

302. 'Most Hens in France Are Still in Cages,' Connexion, March 10, 2020, https://www.connexionfrance.com/news/most-hens-in-france-are-still-in-cages/378865.

303. CIWF, 'Compassion's EggTrack 2024 Spotlights Industry Leaders and Laggards in Cage-free Egg Transition,' November 27, 2024, https://www.ciwf.org/media-news/press-releases-and-media-statements/compassion-s-eggtrack-2024-spotlights-industry-leaders-and-laggards-in-cage-free-egg-transition/.

304. Business Insider Poland, 'Ważna deklaracja Biedronki. Rezygnuje z jaj z chowu klatkowego' [Biedronka's Important Declaration: It Is Phasing Out Cage Eggs], Business Insider, December 3, 2020, https://businessinsider.com.pl/finanse/handel/biedronka-rezygnuje-z-jaj-z-chowu-klatkowego-siec-podala-terminy/z4bj502.

305. Marta Korzeniak, 'Raport: zmiany na polskim rynku jajecznym w 2019 roku'

306. Krajowa Izba Producentów Drobiu i Pasz (KIPDiP), 'Ile klatek w 2026? Prognozy' [How Many Cages in 2026? Forecasts],' Kipdip, March 20, 2018, https://kipdip.org.pl/pl/news/ile-klatek-w-2026-prognozy-.

307. Steve Connor, 'Ban New Battery Cages, Say Activists,' Independent, March 19, 2003, https://www.independent.co.uk/climate-change/news/ban-new-battery-cages-say-activists-111360.htm.

308. 'End the Use of Cages and Crates for All Farmed Animals,' petition, UK Parliament Petitions, https://petition.parliament.uk/petitions/706302; CIWF, 'End the Cage Age,', https://www.ciwf.org.uk/our-campaigns/end-the-cage-age/.

309. European Commission, 'Farm to Fork: Better, but Not Sufficient, Welfare as a Result of EU's Animal Welfare Legislation, Review Finds,' Health and Food Safety Newsroom, October 5, 2022, https://ec.europa.eu/newsroom/sante/items/760589/en.

310. European Commission, 'Tender Details: 7175,' EU Funding & Tenders Portal, https://ec.europa.eu/info/funding-tenders/opportunities/portal/screen/opportunities/tender-details/7175.

311. European Parliament, Annex to Texts Adopted at the Sitting of Wednesday, 23 October 2019: Amendments to the Draft General Budget of the European Union for the Financial Year 2020, Part1/1, Texts Adopted at the Sitting of Wednesday, October 23, 2019, https://www.europarl.europa.eu/cmsdata/188262/P9_TA-PROV(2019)10-23(ANN01)_EN-original.pdf; Best Practice Hens, 'About the Project,', https://bestpracticehens.eu/about-the-project/.

312. European Commission, 'Commission to Propose Phasing Out of Cages for Farm Animals,' press release, June 30, 2021, https://ec.europa.eu/commission/presscorner/detail/en/ip_21_3297.

313. European Commission, Directorate-General for Health and Food Safety, Overview Report on the Protection of the Welfare of Laying Hens at All Stages of Production, (Publications Office, 2023), https://data.europa.eu/doi/10.2875/933391.

314. European Commission, 'European Parliament Hearing: End the Cage Age,' April 15, 2021, https://commission.europa.eu/ec-events/european-parliament-hearing-end-cage-age-2021-04-15_en.

315. Martin Banks, 'European Parliament Calls for the End of All Animal Cages in Europe by 2027,' Parliament Magazine, June 17, 2021, https://www.theparliamentmagazine.eu/news/article/european-parliament-calls-for-the-end-of-all-animal-cages-in-europe-by-2027.

316. Søren Saxmose Nielsen et al., 'Welfare of Laying Hens on Farm,' EFSA Journal 21, no. 2 (February 1, 2023), https://doi.org/10.2903/j.efsa.2023.7789.

317. DeSmog, 'European Livestock Voice,', https://www.desmog.com/european-livestock-voice/.

318. Arthur Neslen, 'Lobby Groups Fought "Hard and Dirty" against EU Ban on Caged Farm Animals,' Guardian, October 23, 2023, https://www.theguardian.com/environment/2023/oct/23/lobby-groups-fought-hard-and-dirty-against-eu-ban-on-caged-farm-animals.

319. Thin Lei Win, 'Citizens vs. Industry: Part II,' Thin Ink, October 18, 2024, https://news.thin-ink.net/p/citizens-vs-industry-part-ii.

320. European Commission, Food, Farming, Fisheries, 'End the Cage Age,' https://food.ec.europa.eu/animals/animal-welfare/eci/eci-end-cage-age_en.

321. Reineke Hameleers, 'Bogus Agricultural Industry Research Threatens to Keep Animals Caged,' Brussels Times, May 29, 2023, https://www.brusselstimes.com/opinion/549178/bogus-agricultural-industry-research-threatens-to-keep-animals-caged.

322. European Commission, Special Eurobarometer 533: Attitudes of Europeans towards Animal Welfare. .

323. European Commission, Proposal for a Regulation of the European Parliament and of the Council on the protection of animals during transport and related operations, Amending Council Regulation (EC) No 1255/97 and Repealing Council Regulation (EC) No 1/2005, COM(2023) 770 final, December 7, 2023, https://food.ec.europa.eu/system/files/2023-12/aw_in-transit_reg-proposal_2023-770_0.pdf

324. EU Policies, Commission Faces Legal Action over Failure to Phase Out Cages in Farming, March 18, 2024, https://eu-policies.com/competences/economy/agriculture/commission-faces-legal-action-failure-phase-cages-farming/.

325. Eurogroup for Animals, 'The Time Is Now – Animal Protection Organisations and MEPs Urge the European Commission to Uphold Its Commitment on Animal Welfare,' press release, June 8, 2023, https://www.eurogroupforanimals.org/news/time-now-animal-protection-organisations-and-meps-urge-european-commission-uphold-its.

326. Sophie Kevany, 'EU-Mercosur Deal Killed Animal Welfare Law,' EUobserver, November 6, 2023, https://euobserver.com/eu-and-the-world/ardb43b97a.

327. European Commission, 'EU and Mercosur Reach Political Agreement on Groundbreaking Partnership,' press release, December 6, 2024, https://ec.europa.eu/commission/presscorner/detail/en/ip_24_6244.

328. DeSmog, 'European Livestock Voice';
Animal Law Europe, 'Industry Lobbying Revealed: Animal Welfare Science Under Siege,' October 18, 2024, https://animallaweurope.org/industry-lobbying-revealed-animal-welfare-science-under-siege/;

Thin Lei Win et al., 'Animal Welfare Wrecked,' Lighthouse Reports, October 23, 2023, https://www.lighthousereports.com/investigation/animal-welfare-wrecked/.

329. Project 1882, 'Sweden Becomes Cage-Free – Project 1882 Celebrates Historic Victory,' June 17, 2025, https://www.project1882.org/news/sweden-becomes-cage-free.

330. STAnews, 'Parliament Passes Reformed Animal Protection Act,' July 18, 2025 'https://english.sta.si/3448923/parliament-passes-reformed-animal-protection-act.

331. UNN, 'Estonia to Ban Keeping Chickens in Cages', January 13, 2025, https://unn.ua/en/news/estonia-to-ban-keeping-chickens-in-cages.

CHAPTER 6: A SECOND CHANCE AT PROGRESS

332. Gregoire Lory and Amandine Hess, 'European Political Landscape Shifts Right in 2024 as Far-Right Gains Ground,' Euronews, December 24, 2024, https://www.euronews.com/my-europe/2024/12/24/european-political-landscape-shifts-right-in-2024-as-far-right-gains-ground.

333. Magnus Lund Nielsen, 'EPP Set to Shape Commissioner Hearings in Power Consolidation Push,' Euractiv, October 31, 2024, https://www.euractiv.com/section/politics/news/epp-set-to-shape-commissioner-hearings-in-power-consolidation-push/.

334. Ursula von der Leyen, Europe's Choice: Political Guidelines for the Next European Commission 2024–2029 (European Commission, July 18, 2024), https://commission.europa.eu/document/download/e6cd4328-673c-4e7a-8683-f63ffb-2cf648_en?filename=Political%20Guidelines%202024-2029_EN.pdf.

335. Ursula von der Leyen, Mission Letter to Olivér Várhelyi, Commissioner-Designate for Health and Animal Welfare, European Commission, September 17, 2024, https://commission.europa.eu/document/download/b1817a1b-e62e-4949-bbb8-ebf29b54c8bd_en?filename=Mission%20letter%20-%20VARHE-LYI.pdf.

336. Ursula von der Leyen, Mission Letter to Christophe Hansen, Commissioner-Designate for Agriculture and Food, European Commission, December 1, 2024, https://www.aaronmcloughlin.com/wp-content/uploads/2024/12/mission-letter-hansen.pdf.

337. European Council, Strategic Agenda 2024–2029, SN 02167/24, June 2024, https://www.consilium.europa.eu/media/yxrc05pz/sn02167en24_web.pdf.

338. Council of the European Union, European Council Conclusions, https://www.consilium.europa.eu/en/european-council/conclusions/.

339. Animal Law Europe,'Brussels at a Standstill: The Fight Over Animal Transport Rules,' May 6, 2025, https://animallaweurope.org/brussels-at-a-standstill-the-fight-over-animal-transport-rules/.

340. Eurogroup for Animals, 'Vote for Animals 2024 | Pledge to Do More!,' April 1, 2025, https://www.eurogroupforanimals.org/vote-animals.

341. Party for the Animals, 'EU Elections: More Than 1.5 Million Votes for the Animals!,' June 24, 2024, https://www.partyfortheanimals.com/es/eu-elections-more-than-1-5-million-votes-for-the-animals.

342. 'Animal Politics EU,' Wikipedia, last modified April 22, 2025, https://en.wikipedia.org/wiki/Animal_Politics_EU.

343. European Commission, Special Eurobarometer 533: Attitudes of Europeans towards Animal Welfare.

344. BEUC, Farm Animal Welfare: What Consumers Want, (February 2024), https://www.beuc.eu/sites/default/files/publications/BEUC-X-2024-016_Farm_animal_welfare_what_consumers_want_survey.pdf.

345. European Commission, Europeans, Agriculture and the CAP, January 2025, https://europa.eu/eurobarometer/surveys/detail/3226.

346. Elena Louazon, 'Belgium Enshrines Animal Welfare in its Constitution,' Le Monde, May 10, 2024, https://www.lemonde.fr/en/environment/article/2024/05/10/belgium-enshrines-animal-welfare-in-its-constitution_6671002_114.html.

347. Federal Ministry of Food and Agriculture, 'Phasing-Out of Chick Culling,', https://www.bmel.de/EN/topics/animals/animal-welfare/research-poultry-in-ovo.html.

348. Ministère de l'Agriculture, de la Viticulture et du Développement rural, 'Une nouvelle loi pour une meilleure protection des animaux' [A New Law for Better Animal Protection],' press release, June 6, 2018, https://deiereschutzgesetz.lu/wp-content/themes/loianimaux/communique-presse.pdf.

349. Natasha Foote, 'Shortsighted Vision: Unpacking EU's New Agrifood Policy Plans,' ARC2020, February 20, 2025, https://www.arc2020.eu/shortsighted-vision-unpacking-eus-new-agrifood-policy-plans/.

350. von der Leyen, Europe's Choice.

351. European Parliament, Verbatim Report of Proceedings, December 19, 2024, CRE-10-2024-12-19, https://www.europarl.europa.eu/doceo/document/CRE-10-2024-12-19_EN.pdf.

352. European Parliament, Briefing: Commitments Made at the Confirmation Hearings of the Commissioners-designate 2024–2029, IPOL_BRI(2025)700896, January 10, 2025, https://www.europarl.europa.eu/RegData/etudes/BRIE/2025/700896/IPOL_BRI(2025)700896_EN.pdf.

353. European Commission, The European Green Deal (December 2019), https://commission.europa.eu/strategy-and-policy/priorities-2019-2024/european-green-deal_en; European Commission, Farm to Fork Strategy, (May 2020), https://food.ec.europa.eu/horizontal-topics/farm-fork-strategy_en.

354. Alessandro Ford, 'The EU Says It Wants Food Security. It Really Wants Exports,' Politico, July 26, 2024, https://www.politico.eu/article/eu-food-security-affordability-exports-farming-agricultural-agenda/.

355. Gerardo Fortuna (@gerardofortuna), 'Is the Farm to Fork dead? 63% of the ongoing initiatives of the EU's flagship food policy are basically dead in the water – namely not proposed yet or even withdrawn – if you see the latest state of play from @EP_ThinkTank,'

356. Gerardo Fortuna, 'That's when I marked the time of death for the Farm to Fork strategy (Commission may still be in denial, but in policymaking, actions matter far more than words),' LinkedIn, February 19, 2025, https://www.linkedin.com/posts/gerardo-fortuna-464237118_1243-19-february-thats-when-i-marked-activity-7336349319660204032-lFqh/.

357. Edward Majewski et al., 'End of the Cage Age? A Study on the Impacts of the Transition from Cages on the EU Laying Hen Sector,' Agriculture 14, no. 1 (2024), https://doi.org/10.3390/agriculture14010111; David Baldock et al., Financing the Cage-Free Farming Transition in Europe (Institute for European Environmental Policy, March 2022), https://ieep.eu/publications/financing-the-cage-free-farming-transition-in-europe/.

358. What is New in the Commission's 2021 Better Regulation Guidelines?, February 2022, https://www.europarl.europa.eu/RegData/etudes/BRIE/2022/699463/EPRS_BRI(2022)699463_EN.pdf.

359. European Commission, 'Eggs – Market Situation – Dashboard.'

360. At the time of writing, Austria, Czechia, Denmark, Germany, Slovenia, and Belgium's Wallonia region have laws phasing out cages, Estonia is considering passing such a law, France has a law restricting new cages from being installed, Slovakia has an agreement from the industry to phase out cages, Sweden and Luxembourg are entirely cage free, and the Netherlands has phased them out for 85% of hens. Finland, Ireland, and Italy are already close to or more than 70% cage free.

361. European Commission, A Vision for Agriculture and Food: Shaping together an attractive farming and agri-food sector for future generations, February 19, 2025, COM(2025) 75 final, https://eur-lex.europa.eu/legal-content/EN/TXT/?uri=CELEX:52025DC0075.

362. European Parliament, Questionnaire to the Commissioner-Designate Olivér Várhelyi, Health and Animal Welfare, (, https://hearings.elections.europa.eu/documents/varhelyi/varhelyi_writtenquestionsandanswers_en.pd.

363. European Parliament, Briefing: Commitments Made.

364. The European Institute for Animal Law & Policy, Recent Developments in EU Animal Law & Policy: 2019 – 2024 in Review (2025), https://animallaweurope.org/wp-content/uploads/Recent-Developments-in-EU-Animal-Law-Policy-2019-2024.pdf.

365. World Trade Organization, Dispute Settlement: DS400 European Communities — Measures Prohibiting the Importation and Marketing of Seal Products, last updated October 16, 2015, https://www.wto.org/english/tratop_e/dispu_e/cases_e/ds400_e.htm.

366. European Commission, Proposal for a Regulation of the European Parliament and of the Council on the welfare of dogs and cats and their traceability, COM(2023) 769 final,December 7, 2023, https://eur-lex.europa.eu/legal-content/EN/TXT/?uri=celex:52023PC0769; The European Institute for Animal Law & Policy, 'June 2025 Newsletter – Tracking Progress: Cats and Dogs Regulation Advances in EU,' Animal Law Europe, June 11, 2025, https://animallaweurope.substack.com/p/june-2025-newsletter-tracking-progress.

CHAPTER 7: FARMED FISH WELFARE: THE NEXT FRONTIER FOR EU LEADERSHIP

367. Eurogroup for Animals, 'Seizing the Day for Fish Welfare 6,' YouTube, June 20, 2018, https://www.youtube.com/watch?v=Np89ndGZmj4;

Fishcount, 'Estimated Numbers of Individuals in Aquaculture Production (FAO) of Fish Species (2017):

368. Mood et al., 'Estimating Global Numbers of Farmed Fishes Killed for Food.'

369. Hannah Ritchie et al., 'Meat and Dairy Production,' Our World in Data, last updated December 2023, https://ourworldindata.org/meat-production.

370. M.N.J. Turenhout et al., EU Seafood Supply Synopsis 2024 (AIPCE-CEP, October 2024), https://www.aipce-cep.org/wp-content/uploads/2024/10/EU-Seafood-Supply-Synopsis-_2024.pdf.

371. European Commission, Directorate-General for Maritime Affairs and Fisheries, 'Overview of EU Aquaculture (Fish Farming),', https://oceans-and-fisheries.ec.europa.eu/ocean/blue-economy/aquaculture/overview-eu-aquaculture-fish-farming_en;
European Climate, Infrastructure and Executive Agency, 'EU Aquaculture Sector: Socioeconomic Development (2008–2020)' August 2023, https://aquaculture.ec.europa.eu/system/files/2023-08/EU%20Aquaculture%20Sector_Socioeconomic%20development_Infographic.pdf;
Mood et al., 'Estimating Global Numbers of Farmed Fishes Killed for Food.'

372. European Commission, Overview of EU Aquaculture;
Eurofish International Organisation, 'Overview of the Croatian Fisheries and Aquaculture Sector,' December 14, 2023, https://eurofish.dk/member-countries/croatia/;
Food and Agriculture Organization of the United Nations, 'National Aquaculture Sector Overview: Cyprus,', https://www.fao.org/fishery/en/countrysector/cy/en. See also farm animal statistics cited in Lewis Bollard, 'Why Did the EU Lead the World on Farm Animal Welfare, and How Can It Lead Again?,' Open Philanthropy, September 18, 2017,

373. European Market Observatory for Fisheries and Aquaculture Products (EUMOFA), Large Trout in the EU, Price Structure in the Supply Chain;
Focus on Spain and Italy (Publications Office of the European Office, 2023), 19, https://eumofa.eu/documents/20178/543766/PTAT_Large+trout.pdf;
EUMOFA, Fresh Gilthead Seabream in the EU, Price Structure in the Supply Chain: Focus on Spain, Germany and France (Publications Office of the European Union, 2022), 10, https://eumofa.eu/documents/20178/486475/PTAT+-fresh+seabream+in+ES+FR+and+DE_EN.pdf;
EUMOFA, Seabass in the EU, Price Structure in the Supply Chain: Focus on Greece, Croatia and Spain (Publications Office of the European Union, 2021 [revised in January 2019]), https://eumofa.eu/documents/20178/121372/PTAT+-Case+Study+-+Seabass+in+the+EU.pdf.

374. Neil Dullaghan, EU Farmed Fish Policy Reform Roadmap Brief (Rethink Priorities, August 21, 2023), https://rethinkpriorities.org/research-area/eu-farmed-fish-policy-reform-roadmap-brief/.

375. Hellenic Aquaculture Producers Organization, Greek Aquaculture 2022 Annual Report (Hellenic Aquaculture Producers Organization, 2022), 79, 82, https://fishfromgreece.com/wp-content/flipbook/nov22/.

376. Lela Nargi, 'Fish Make Sounds That Could Help Scientists Protect Them,' Washington Post, May 10, 2022, https://www.washingtonpost.com/kids-post/2022/05/10/fish-make-sounds-that-could-help-scientists-protect-them/.

377. Jonathan A.C. Roques et al., 'Tailfin Clipping, a Painful Procedure: Studies on Nile Tilapia and Common Carp,' Physiology & Behavior 101, no. 4 (November 2, 2010): 533–40, https://www.sciencedirect.com/science/article/abs/pii/S0031938410002866?via%3Dihub.

378. K.P. Chandroo et al., 'Can Fish Suffer? Perspectives on Sentience, Pain, Fear and Stress,' Applied Animal Behaviour Science 86, no. 3 (June 2004): 225–50, https://www.sciencedirect.com/science/article/abs/pii/S0168159104000498.

379. Lynne U. Sneddon et al., 'Do Fishes Have Nociceptors? Evidence for the Evolution of a Vertebrate Sensory System,' Proceedings of the Royal Society B: Biological Sciences 270, no. 1520 (June 7, 2003): 1115–21, https://royalsocietypublishing.org/doi/10.1098/rspb.2003.2349.

380. Lynne U. Sneddon, 'Evolution of Nociception and Pain: Evidence from Fish Models,' Philosophical Transactions of the Royal Society B: Biological Sciences 374, no. 1785, art. 20190290 (November 11, 2019): 20190290, https://pubmed.ncbi.nlm.nih.gov/31544617/.

381. Culum Brown, 'Fish Intelligence, Sentience and Ethics,' Animal Cognition 18, no. 1 (2015): 1–17, https://www.wellbeingintlstudiesrepository.org/cgi/viewcontent.cgi?article=1074.

382. European Commission, Answer to Written Parliamentary Question E-1140/2009, April 3, 2009, http://www.europarl.europa.eu/sides/getAllAnswers.do?reference=E-2009-1140.

383. Council of the European Union, Council Regulation (EC) No 1/2005 of 22 December 2004 on the protection of animals during transport and related operations and amending Directives 64/432/EEC and 93/119/EC and Regulation (EC) No 1255/97, Official Journal of the European Union L 3, January 5, 2005, 1–44, https://eur-lex.europa.eu/eli/reg/2005/1/oj/eng;
Alice Di Concetto et al., The Treatment of Farmed Fish Under EU Law (The European Institute for Animal Law & Policy, 2024), https://animallaweurope.org/wp-content/uploads/Research-Note-7_The-Treatment-of-Farmed-Fish-Under-EU-Law.pdf; Marita Giménez-Candela et al., 'Legal Protection of Farmed Fish in Europe: Analysis of EU Legislation and the Impact of International Animal Welfare Standards on Farmed Fish in Europe,' Derecho Animal, Forum of Animal Law Studies 11, no. 1 (2020), https://revistes.uab.cat/da/article/view/v11-n1-gimenez-candela-saraiva-bauer.

384. EFSA, 'Opinion of the Scientific Panel on Animal Health and Welfare on a Request from the Commission Related to Welfare Aspects of the Main Systems of Stunning and Killing the Main Commercial Species of Animals,' EFSA Journal 45 (July 6, 2004): 1–29, https://doi.org/10.2903/j.efsa.2004.45.

385. Saulius Šimčikas, 'How Much Do Europeans Care About Fish Welfare? (An Analysis of Relevant Surveys),' Effective Altruism Forum, June 22, 2020, https://forum.effectivealtruism.org/posts/wDGAvyTafuqFznqko/how-much-europeans-care-about-fish-welfare.

386. European Commission, 'EU Fish Farms (Aquaculture): Updated Guidelines – Public Consultation,', https://ec.europa.eu/info/law/better-regulation/have-your-say/initiatives/12261-EU-fish-farms-aquaculture-updated-guidelines/public-consultation_en.

387. Eurogroup for Animals, Looking Beneath the Surface: Fish Welfare in European Aquaculture, (Version 2, July 2018), https://web.archive.org/web/20201124083810/https:/www.eurogroupforanimals.org/sites/eurogroup/files/2020-02/Fish-Welfare-in-European-Aquaculture-2.pdf.

388. European Commission, EU Animal Welfare Strategy (2012-15) – evaluation, https://ec.europa.eu/info/law/better-regulation/have-your-say/initiatives/2140-EU-animal-welfare-strategy-2012-15-evaluation_en.

389. Anne Altmayer, 'EU Aquaculture: State of Play,' European Parliamentary Research Service (EPRS), Members' Research Service, June 2024, https://www.europarl.europa.eu/thinktank/en/document/EPRS_BRI%282024%29762336.

390. Council on Animal Affairs (RDA), Fish Welfare (RDA, March 7, 2018), https://english.rda.nl/documents/2018/03/07/fish-welfare.

391. Michalis Pavlidis and Antonios Samaras, Mediterranean Fish Welfare: Guide to Good Practices and Assessment Indicators (Hellenic Aquaculture Producers Organization and University of Crete, December 2019), https://aquaculture.ec.europa.eu/system/files/2023-08/HAPO%20mediterranean%20fish%20welfare%20guide.pdf.

392. Carpendale and Bridgwater, 'Mapping the industry and supply chain for farmed fish in Europe.' (Animal Ask, May 2025) https://www.animalask.org/post/mapping-the-industry-and-supply-chain-for-farmed-fish-in-europe

393. Ren Ryba, The Economics of Fish Farming and Fish Welfare in Europe: A Systematic Review (Animal Ask, February 2025, updated May 5), https://www.animalask.org/post/the-economics-of-fish-farming-and-fish-welfare-in-europe#viewer-ttx5n33896;
EUMOFA, The EU Fish Market, 2023 edition (Publications Office of the European Union), https://eumofa.eu/documents/20124/35668/EFM2023_EN.pdf.

394. EFSA, 'Opinion of the Scientific Panel on Animal Health and Welfare.'

395. Council Regulation (EC) No 1099/2009 of 24 September 2009 on the protection of animals at the time of killing, Official Journal of the European Union L 303, November 18, 2009, 1, https://eur-lex.europa.eu/LexUriServ/LexUriServ.do?uri=OJ%3AL%3A2009%3A303%3A0001%3A0030%3AEN%3APDF.

396. EFSA, 'Fish welfare,', https://www.efsa.europa.eu/en/topics/topic/fish-welfare.

397. 'About Us: What is Aquaculture?', Aquaculture Advisory Council, https://aac-europe.org/en/about-us/what-is-aquaculture/.

398. European Commission: Directorate-General for Health and Food Safety, VetEffecT, Wageningen University, K. van der Braak, R. Schrijver, et al., Welfare of Farmed Fish – Common Practices During Transport and at Slaughter – Final report (Publications Office, 2017), https://data.europa.eu/doi/10.2875/172078.

399. European Parliament, Animal welfare rules in aquaculture (debate), Verbatim Report of Proceedings, March 14, 2019, https://www.europarl.europa.eu/doceo/document/CRE-8-2019-03-14-ITM-016_EN.html

400. Council of the European Union, Council conclusions on animal welfare – an integral part of sustainable animal production, 14975/19 (Brussels, December 16, 2019), https://www.consilium.europa.eu/media/41863/st14975-en19.pdf.

401. EU Platform on Animal Welfare Own Initiative Group on Fish, Guidelines on Water Quality and Handling for the Welfare of Farmed Vertebrate Fish,

DOC.11068.202.Rev.1, 2020, https://drive.google.com/file/d/1L17_bPtpsD-pHA8OaDWbwEb3WJMbUhJ3N/view.

402. European Parliament, Multimedia Centre, 'Committee on Fisheries,' video, https://multimedia.europarl.europa.eu/en/webstreaming/committee-on-fisheries_20250127-1500-COMMITTEE-PECH.

403. European Commission, Commission Implementing Regulation (EU) 2020/464 of 26 March 2020 laying down certain rules for the application of Regulation (EU) 2018/848 concerning organic production and labeling of organic products, Official Journal of the European Union L 98, March 31, 2020, 2–25, https://eur-lex.europa.eu/legal-content/EN/TXT/PDF/?uri=CELEX:32020R0464.

404. EUMOFA, The EU Fish Market.

405. World Organisation for Animal Health (OIE), Aquatic Animal Health Strategy 2021–2025 (World Organisation for Animal Health [OIE], May 2021), https://www.woah.org/app/uploads/2021/05/en-oie-aahs.pdf.

406. EU Platform on Animal Welfare Own Initiative Group on Fish, Guidelines on Water Quality and Handling for the Welfare of Farmed Vertebrate Fish.

407. Council of Europe, Standing Committee on the European Convention for the PRotection of Animals Kept for Farming Purposes (T-AP) Recommendation Concerning Farmed Fish, (Rec Fish E), entered into force June 5, 2006, https://www.coe.int/t/e/legal_affairs/legal_co-operation/biological_safety_and_use_of_animals/Farming/Rec%20fish%20E.asp.

408. Ministry of Fisheries and Coastal Affairs (Norway), Forskrift om slakterier og tilvirkingsanlegg for akvakulturdyr [Regulation on slaughterhouses and processing plants for aquaculture animals], FOR-2006-10-30-1250, October 30, 2006, effective January 1, 2007, Lovdata, https://lovdata.no/dokument/SF/forskrift/2006-10-30-1250/KAPITTEL_4#%C2%A714.

409. The Swiss Federal Council, Animal Welfare Ordinance (TSchV; SR 455.1), April 23, 2008, https://www.globalanimallaw.org/downloads/database/national/switzerland/TSchV-2008-EN-455.1-2011.pdf.

410. Bundesministerium der Justiz, Verordnung zum Schutz von Tieren im Zusammenhang mit der Schlachtung oder Tötung und zur Durchführung der Verordnung (EG) Nr. 1099/2009 (Tierschutz-Schlachtverordnung, TierSchlV 2013) [Regulation on the protection of animals in connection with slaughter or killing and implementing Council Regulation (EC) No. 1099/2009 (Animal Welfare and Slaughter Regulation - TierSchlV)], Kap. 4, § 14, promulgated December 20, 2012, BGBl I 2012, 2982, effective January 1, 2013, https://www.gesetze-im-internet.de/tierschlv_2013/BJNR298200012.html.

411. Ministry for Primary Industries (New Zealand), Code of Welfare:Commercial Slaughter (Ministry for Primary Industries, April 27, 2021; last amended March 21, 2025), https://www.mpi.govt.nz/dmsdocument/46018-Code-of-Welfare-Commercial-slaughter.

412. Chris Chase, 'Maine Governor Janet Mills Signs Bill Placing Limits on Salmonid Aquaculture,' SeafoodSource, June 26, 2023, https://www.seafoodsource.com/news/aquaculture/maine-passes-bill-placing-limits-on-salmon-aquaculture.

413. Ministry of Fisheries and Coastal Affairs (Norway) , Forskrift om drift av akvakulturanlegg [Regulation on the operation of aquaculture facilities],' FOR-2008-06-17-822, Kapittel 3, § 25, June 17, 2008, last amended February 2, 2024,

Lovdata, https://lovdata.no/dokument/SF/forskrift/2008-06-17-822/KAPIT-TEL_3#%C2%A725.

414. Swiss Federal Council, Animal Welfare Ordinance.

415. Christian Perez Mallea, 'New Regulation on Stocking Densities: Looking at the Individual and Collective Performance,' Fish Farming Expert, June 19, 2015 (modified February 24, 2018), https://www.fishfarmingexpert.com/new-regu-lation-on-stocking-densities-looking-at-the-individual-and-collective-perfor-mance/1293817.

416. Swiss Federal Council, Animal Welfare Ordinance.

417. Ministry of Agriculture (Czech Republic), Vyhláška č. 418/2012 Sb. o ochraně zvířat při usmrcování [Decree No. 418/2012 Coll. on the Protection of Animals During Killing], Kap. 3, § 7, promulgated November 22, 2012, effective January 1, 2013, Sbírka zákonů, https://www.zakonyprolidi.cz/nabidka/cs/2012-418/zne-ni-20130101#f4825949_p9-2.

418. Food and Agriculture Organisation of the United Nations (FAO), Republic of Tur-key, Regulation on fishing vessel safety No. 22223, enacted 22 July 1995, FAOLEX Database, http://www.fao.org/faolex/results/details/en/c/LEX-FAOC170083/.

419. Sagar Shah, Prospective Cost-Effectiveness of Farmed Fish Stunning Corporate Commitments in Europe, (Rethink Priorities, March 14, 2024), https://rethink-priorities.org/research-area/farmed-fish-corporate-commitments/; Carpendale and Bridgwater, Mapping the Industry and Supply Chain for Farmed Fish in Europe, (Animal Ask, May 2025), https://www.animalask.org/post/mapping-the-industry-and-supply-chain-for-farmed-fish-in-europe.

420. Ace Aquatec, 'Exploring New Areas and Species for Ace Aquatec Stunners,' Jan-uary 18, 2023, https://aceaquatec.com/news-and-resources/news/exploring-new-areas-and-species-ace-aquatec-stunners;
Ace Aquatec, 'Greek Aquaculture Company, Philosofish, Leads the Way in Fish Welfare with Installation of Ace Aquatec Humane Stunners,' August 9, 2023, https://aceaquatec.com/news-and-resources/news/greek-aquaculture-compa-ny-philosofish-leads-way-fish-welfare-installation-ace-aquatec-humane-stun-ners;
'Ace Aquatec to Bring Their Electric Stunners to Greece,' The Fish Site, October 18, 2022, https://thefishsite.com/articles/ace-aquatech-to-bring-their-electric-stunners-to-greece;
Natasha Boyland, The Welfare of Farmed Fish during Slaughter in the Europe-an Union (Compassion in World Farming, November 2018), https://www.ciwf. org.uk/media/7434891/ciwf-2018-report__the-welfare-of-farmed-fish-during-slaughter-in-the-eu.pdf;
Pricillia Durbant et al., Welfarm Farmed Fish Slaughter Methods Report (Wel-farm – Protection mondiale des animaux de ferme, March 2023), 70–72, https:// welfarm.fr/wp-content/uploads/2023/03/WELFARM-HUMANE-SLAUGH-TER-FOR-FARMED-FISH.pdf.

421. Aquaculture Stewardship Council, 'ASC Position Statement on Stunning,' last updated November 12, 2024, https://asc-aqua.org/asc-position-state-ment-on-stunning/.

422. Commission Regulation (EC) No 710/2009 of 5 August 2009 amending Reg-ulation (EC) No 889/2008 laying down detailed rules for the implementation of Council Regulation (EC) No 834/2007 with regard to organic aquaculture

animal and seaweed production, Official Journal of the European Union L 204, 6 August 2009, 15–34, https://eur-lex.europa.eu/legal-content/EN/TXT/PD-F/?uri=CELEX:32009R0710.

423. Ren Springlea, Economic Evaluation of Humane Slaughter Methods for Farmed Fish in Greece (Eurogroup for Animals, April 2022), https://www.eurogroupfor-animals.org/files/eurogroupforanimals/2023-02/Greece_Humane%20Slaughter%20for%20Farmed%20Fish.pdf;
Hans van de Vis et al., Welfare of Farmed Fish: Common Practices during Transport and at Slaughter (Publications Office of the European Union, 2017), https://publications.europa.eu/resource/cellar/facddd32-cda6-11e7-a5d5-01aa75ed71a1.0001.01/DOC_1.

424. Michail Pavlidis et al., Animal Welfare of Farmed Fish (European Parliament Policy Department for Structural and Cohesion Policies, June 2023), https://www.europarl.europa.eu/RegData/etudes/STUD/2023/747257/IPOL_STU%282023%29747257_EN.pdf.

425. Henri Prins et al., OrAqua Deliverable 3.2: Farm Economics and Competitiveness of Organic Aquaculture (LEI Wageningen UR, 2023), https://www.oraqua.eu/content/download/110481/file/OrAqua%20D%203_2.pdf;
European Commission, Commission Implementing Regulation (EU) 2020/464 of March 2020.

426. Pavlidis and Samaras, Mediterranean Fish Welfare https://aquaculture.ec.europa.eu/system/files/2023-08/HAPO%20mediterranean%20fish%20welfare%20guide.pdf

427. CIWF, Improving the Welfare of European Sea Bass and Gilthead Sea Bream, https://www.compassioninfoodbusiness.com/media/7436996/the-science-driving-change-for-gilthead-sea-bream-and-european-sea-bass.pdf.

428. Pavlidis and Samaras, Mediterranean Fish Welfare.

429. CIWF, Improving the Welfare of European Sea Bass and Gilthead Sea Bream.

430. Reet Kaur, 'Record Marine Heatwave Pushes Mediterranean Sea Surface Temperatures to 30°C (86°F) Off Spain,' The Watchers, July 2, 2025, https://watchers.news/2025/07/02/record-marine-heatwave-mediterranean-sea-surface-temperatures-30c-spain/.

431. Swiss Federal Council, Animal Welfare Ordinance.

432. Prins et al., OrAqua Deliverable 3.2.

433. Di Concetto et al., The Treatment of Farmed Fish Under EU Law.

CHAPTER 8: OVERCOMING THE BARRIERS TO PROGRESS

434. Council of the European Union, Council Directive 2008/119/EC of 18 December 2008 laying down minimum standards for the protection of calves (codified version), Official Journal of the European Union L 10, January 15, 2009, 7–13, https://eur-lex.europa.eu/legal-content/EN/ALL/?uri=CELEX:32008L0119;
Council of the European Union, Council Directive 2008/120/EC of 18 December 2008 laying down minimum standards for the protection of pigs, Official Journal of the European Union L 47, February 18, 2009, 5–13, https://eur-lex.europa.eu/legal-content/EN/TXT/?uri=CELEX:32008L0120.

435. Council of the European Union, Council Directive 1999/74/EC of 19 July 1999 laying down minimum standards for the protection of laying hens, Official Journal of the European Union L 203, August 3, 1999, 53–57, https://eur-lex. europa.eu/legal-content/EN/TXT/?uri=CELEX:31999L0074.

436. European Union, Rules on Marketing Standards for Eggs (Summary of Regulation (EC) No 589/2008 of 23 June 2008), EUR-Lex Summaries of EU Legislation, last updated February 27, 2018, https://eur-lex.europa.eu/legal-content/EN/TXT/?uri=LEGISSUM:4324376.

437. Council of the European Union, Council Directive 2007/43/EC of 28 June 2007 laying down minimum rules for the protection of chickens kept for meat production, Official Journal of the European Union L 182, July 12, 2007, 19–28, https://eur-lex.europa.eu/legal-content/EN/TXT/?uri=CELEX:32007L0043.

438. Fur Free Alliance, 'Fur Bans,', https://www.furfreealliance.com/fur-bans/.

439. European Union, Fur Free Europe: European Citizens' Initiative (ECI-2022/000002) aiming for an EU-wide ban on keeping and killing of animals for the sole or main purpose of fur production, registered 16 March 2022, Citizens' Initiative (European Union), https://citizens-initiative.europa.eu/initiatives/details/2022/000002_en.

440. Service Public Fédéral Justice (Belgium), Arrêté du 22 décembre 2023 fixant les prescriptions minimales pour la protection des dindons [Order of December 22, 2023 setting the minimum requirements for the protection of turkeys], Moniteur belge, January 24, 2024, https://www.ejustice.just.fgov.be/cgi/article_body.pl?language=fr&caller=summary&pub_date=24-01-24&numac=2024000463.

441. Eurogroup for Animals, 'Olympic Games to Serve Foie Gras: Stark Reminder This Cruel Industry Continues,' February 21, 2023, https://www.eurogroupforanimals.org/news/olympic-games-serve-foie-gras-stark-reminder-cruel-industry-continues.

442. For more information, see the European Institute for Animal Law & Policy, Recent Developments in EU Animal Law & Policy: 2019 – 2024 in Review (2025), https://animallaweurope.org/wp-content/uploads/Recent-Developments-in-EU-Animal-Law-Policy-2019-2024.pdf.

443. For more information, see Alice Di Concetto et al., Animal Welfare Standards in EU Organic Certification, (The European Institute for Animal Law & Policy, 2022), https://animallaweurope.org/wp-content/uploads/Research-Note-5-Animal-Welfare-Standards-in-the-EU-Organic-Certification-1.pdf.

444. Bartosz Brzeziński and Paula Andrés, 'Meet the Farmers Who Control One-Third of the EU's Budget,' Politico, October 6, 2023, https://www.politico.eu/article/eu-parliament-farmers-who-control-one-third-of-the-eu-budget-franc-bogovic/;
Lewis Bollard, 'This Is Why We Can't Have Nice Laws,' Open Philanthropy Farm Animal Welfare Newsletter, February 27, 2024, https://farmanimalwelfare.substack.com/p/this-is-why-we-cant-have-nice-laws.

445. CIWF, 'MEPs Fail Citizens on Animal Welfare, New Analysis Reveals,' October 13, 2023, https://www.ciwf.eu/media-and-news/news/meps-fail-citizens-on-animal-welfare-new-analysis-reveals/.

446. Eurogroup for Animals, 'Setting the Pace for Another Four Decades Making Headway for Animals,' February 28, 2020, https://www.eurogroupforanimals. org/news/setting-pace-another-four-decades-making-headway-animals.

447. Environmental Working Group, 'Despite Record Farm Income and Subsidies, Some Seek Even More Handouts,' September 2023, https://www.ewg. org/news-insights/news/2023/09/despite-record-farm-income-and-subsidies-some-seek-even-more-handouts.

448. European Livestock Voice, https://meatthefacts.eu/european-livestock-voice/.

449. Susannah Savage and Thin Lei Win, 'The Truth Behind Europe's Most Powerful Farmers Lobby,' Politico, June 29, 2023, https://www.politico.eu/article/ copa-cogeca-farmering-lobby-europe/.

450. Corporate Europe Observatory, 'Far Right Aims to Capture Farmer Protests – But Offers No Solutions, Promoting Same Neoliberal Agenda,' June 4, 2024, https://corporateeurope.org/en/2024/06/far-right-aims-capture-farmer-protests-offers-no-solutions-promoting-same-neoliberal-agenda.

451. European Forum of Farm Animal Breeders (EFFAB), COPA-COGECA, and AVEC, 'Joint Statement, EFSA Scientific Opinion on Broilers on Farms: The Roadmap to Unsustainable Poultry Production in Europe,' February 21, 2023, https://web.archive.org/web/20230604042352/https://www.effab.info/wp-content/uploads/2023/03/Final-joint-statement-with-COPA-and-AVEC.pdf.

452. Harrison, Animal Machines.

453. Nicholas Kristof, 'The Truth About Your Bacon,' New York Times, August 5, 2023, https://www.nytimes.com/2023/08/05/opinion/hog-farming-secret-video.html.

454. European Commission, Commission Work Programme, https://commission. europa.eu/strategy-and-policy/strategy-documents/commission-work-programme_en.

455. Nicholas Kristof, 'The Ugly Secrets Behind the Costco Chicken,' New York Times, February 6, 2021, https://www.nytimes.com/2021/02/06/opinion/sunday/costco-chicken-animal-welfare.html.

456. Ezra Klein, 'Farmers and Animal Rights Activists Are Coming Together to Fight Big Factory Farms,' Vox, July 8, 2020, https://www.vox.com/future-perfect/2020/7/8/21311327/farmers-factory-farms-cafos-animal-rights-booker-warren-khanna

www.ingramcontent.com/pod-product-compliance
Lightning Source LLC
Chambersburg PA
CBHW052020030426
42335CB00026B/3225